Mathematical Wisdom in Everyday Life

From Common Core to Math Competitions

Solutions Manual

Kevin Wang
Kelly Ren
John Lensmire

PUBLISHED BY ARETEEM INSTITUTE

WWW.ARETEEM.ORG

ISBN: 1-944863-04-4
ISBN-13: 978-1-944863-04-3

First printing, September 2016.

Contents

Introduction

This book is a part of the ongoing effort by Areteem Institute to inspire students, parents, and teachers to gain a deeper understanding and appreciation of mathematics. The content is organized to emphasize on the proper implementation of the Common Core Mathematics Standard, focusing on conceptual understanding, problem solving, and real world applications. Some contest (AMC8, MATHCOUNTS, and ZIML Division M) level problems are included to challlenge talented middle school students or advanced elementary school students.

This book contains complete solutions of the problems in the book "*Mathematical Wisdom in Everyday Life*". The key concepts, examples and their solutions, and problem solving strategies can be found in the book "*Mathematical Wisdom in Everyday Life*".

This book contains many years of collaborative work by the staff of Areteem Institute. This book could not have existed without their efforts. Huge thanks go to the Areteem staff for their contributions.

The problems in this book were either created by the Areteem staff or adapted from various sources, including other books and online resources. It is not practical to list all such resources here. We extend our gratitude to the original authors of all these resources.

About Areteem Institute

Areteem Institute is an educational institution that develops and provides in-depth and advanced math and science programs for K-14 (Elementary School, Middle School, High School, and two years of college) students and teachers. Areteem programs are accredited supplementary programs by the Western Association of Schools and Colleges (WASC). Students may attend the Areteem Institute through these options:

- Live and real-time face-to-face online classes;
- Self-paced classes by watching the recordings of the live classes;
- Summer Intensive Camps and Winter Boot Camps.

The Areteem courses are designed and developed by educational experts and industry professionals to bring real world applications into the STEM education. The programs are ideal for students who wish to win in Math Competitions (AMC, AIME, USAMO, IMO, ARML, MathCounts, Math Olympiad, ZIML, etc.), Science Fairs (County Science Fairs, State Science Fairs, national programs like Intel Science and Engineering Fair, etc.) and Science Olympiad, or purely want to enrich their academic lives by taking more challenges and developing outstanding analytical, logical thinking and creative problem solving skills.

Since 2004 Areteem Institute has been teaching with the methodology that is highly promoted by the new Common Core State Standards: stressing the conceptual level understanding of the math concepts, problem solving techniques, and solving problems in the real world applications. With the guidance from experienced and passionate professors, students are motivated to explore concepts deeper by identifying an interesting problem, researching it, analyzing it, and using a critical thinking approach to come up with multiple solutions.

Hundreds of math students who have been trained at Areteem achieved top honors and earned top awards in major national and international math competitions, including Gold Medalists in the International Math Olympiad (IMO), top winners at the USA Math Olympiad (USAMO/JMO), top winners at the Zoom International Math League (ZIML), and top winners at the MathCounts National. Many Areteem Alumni have graduated from high school and gone on to enter their dream colleges such as MIT, Cal Tech, Harvard, Stanford, Yale, Princeton, U Penn, Harvey Mudd College, UC Berkeley, UCLA, etc. Those who have graduated from college are now playing important roles in their fields of endeavor.

Further information about Areteem Institute, as well as updates and errata of this book, can be found online at http://www.areteem.org.

1. Number Sense

Problem 1.1 **Find all the factors of 15.**

Answer

$1, 3, 5, 15$

Solution

The only ways to write 15 as the product of two numbers is

$$15 = 1 \times 15 = 15 \times 1 \text{ or } 15 = 3 \times 5 = 5 \times 3.$$

Therefore the factors of 15 are $1, 3, 5, 15$.

Problem 1.2 **How many factors does the number 30 have?**

Answer

8

Solution

We can write

$$30 = 1 \times 30 = 2 \times 15 = 3 \times 10 = 5 \times 6.$$

Since factors come in pairs, we know that 30 has factors $1, 30, 2, 15, 3, 10, 5, 6$. Counting there are a total of 8 different factors.

Problem 1.3 **How many factors does the number 16 have?**

Answer

5

Solution

We can write

$$16 = 1 \times 16 = 2 \times 8 = 4 \times 4$$

Since factors come in pairs, we know that 16 has factors $1, 16, 2, 8, 4, 4$. Since we only count 4 as a factor once, there are a total of 5 factors.

Problem 1.4 **List all the factors of 120 less than or equal to 10.**

Answer

$1, 2, 3, 4, 5, 6, 8, 10.$

Solution

It is helpful to first note that

$$120 = 10 \times 12$$

We can then list the factors of 10, which are $1, 2, 5, 10$ and of 12, which are $1, 2, 3, 4, 6, 10$. Each of these are factors of 120, giving us 1, 2, 3, 4, 5, 6, and 10. Further, the combinations of these also give us factors. The only new combination is $2 \times 4 = 8$. Hence, the numbers

$$1, 2, 3, 4, 5, 6, 8, 10$$

are the factors of 120 less than 10.

Problem 1.5 Simplify the fraction $\dfrac{3}{36}$.

Answer

$$\frac{1}{12}$$

Solution

First note that 3 is a factor of 36 because $36 \div 3 = 12$. Therefore we can divide the numerator and denominator by 3 to get

$$\frac{3}{36} = \frac{1}{12},$$

so $\dfrac{1}{12}$ is the simplified fraction.

Problem 1.6 Find the quotient and remainder when $35 \div 3$.

Answer

Quotient: 11, Remainder: 3

Solution

Notice that 33 is the largest multiple of 3 below 35. Since

$$33 = 11 \times 3$$

we have the quotient is 11. This leaves

$$35 - 33 = 2$$

left over, so the remainder is 2.

Problem 1.7 Write the improper fraction $\dfrac{41}{4}$ as a mixed number.

Answer

$$10\frac{1}{4}$$

Solution

We first notice that
$$40 = 10 \times 4$$
is the largest multiple of 4 less than 41. This leaves a remainder of
$$41 - 40 = 1$$
so we have
$$\frac{41}{4} = 10 + \frac{1}{4} = 10\frac{1}{4}$$
when written as a mixed number.

Problem 1.8 Write the improper fraction $\dfrac{39}{6}$ as a simplified mixed number.

Answer

$6\dfrac{1}{2}$

Solution

Notice first that
$$6 \times 6 = 36$$
is the largest multiple of 6 less than 39. Therefore, $39 \div 6$ is 6 with remainder 3. Thus,
$$\frac{39}{6} = 6 + \frac{3}{6} = 6\frac{3}{6}.$$
Do not forget to check your answer is simplified! 3 is a factor of 6, so we can write
$$\frac{3}{6} = \frac{3 \div 3}{6 \div 3} = \frac{1}{2}.$$
Therefore,
$$\frac{39}{6} = 6\frac{1}{2}$$
when written as a simplified mixed number.

Problem 1.9 Write the mixed number $4\dfrac{5}{7}$ as an improper fraction.

Answer

$$\frac{33}{7}$$

Solution

We have that $4 \times 7 = 28$, so

$$4 = \frac{28}{7}.$$

Therefore,

$$4\frac{5}{7} = 4 + \frac{5}{7} = \frac{28}{7} + \frac{5}{7} = \frac{28+5}{7} = \frac{33}{7}$$

so $4\frac{5}{7}$ as an improper fraction is $\frac{33}{7}$.

Problem 1.10   **What is $\dfrac{8}{19} + \dfrac{11}{19}$?**

Answer

1

Solution

Since the fractions have the same denominator we add the numerators, so

$$\frac{8}{19} + \frac{11}{19} = \frac{8+11}{19} = \frac{19}{19}.$$

Simplifying,

$$\frac{19}{19} = 1$$

so our final answer is 1.

Problem 1.11   **Calculate $\dfrac{5}{24} + \dfrac{7}{24}$. Write your answer as a simplified fraction.**

Answer

$$\frac{1}{2}$$

Solution

Since the two fractions have the same denominator, we add the numerators.

$$\frac{5}{24} + \frac{7}{24} = \frac{5+7}{24} = \frac{12}{24}.$$

Then note that 12 is a factor of 24, so we can simplify our answer. Therefore, in simplified form

$$\frac{12}{24} = \frac{12 \div 12}{24 \div 12} = \frac{1}{2}$$

is our final answer.

Problem 1.12   **Calculate** $6 - \frac{7}{3}$**. Write your answer as an improper fraction.**

Answer

$$\frac{11}{3}$$

Solution

We have $6 \times 3 = 18$, so

$$6 = \frac{18}{3}.$$

Then since we are subtracting two fractions with the same denominator we subtract the numerators so

$$\frac{18}{3} - \frac{7}{3} = \frac{11}{3},$$

so our answer is $\frac{11}{3}$ when written as an improper fraction.

Problem 1.13   **What is** $\frac{15}{11} + \frac{19}{11}$**? Write your answer as a mixed number.**

Answer

$$3\frac{1}{11}$$

Solution

Both fractions have the same denominator, so we add numerators to get

$$\frac{15}{11} + \frac{19}{11} = \frac{34}{11}.$$

Then note $34 \div 11$ is 3 with remainder 1, so

$$\frac{34}{11} = 3 + \frac{1}{11} = 3\frac{1}{11}$$

as our final answer.

Problem 1.14 **What is the sum of the mixed numbers $2\frac{1}{8}$ and $4\frac{3}{8}$? Write your answer as a mixed number with a simplified fraction.**

Answer

$$6\frac{1}{2}$$

Solution

We add the whole numbers and fractions separately. We have

$$2 + 4 = 6.$$

The fractions have the same denominator, so we add the numerators to get

$$\frac{1}{8} + \frac{3}{8} = \frac{4}{8};$$

since 4 is a factor of 8, we divide the numerator and denominator by 4 to get

$$\frac{4}{8} = \frac{4 \div 4}{8 \div 4} = \frac{1}{2}.$$

Combining the whole number part with the fraction part we have

$$6\frac{1}{2}$$

as our final answer.

Problem 1.15 Multiply the fractions $\dfrac{3}{5} \times \dfrac{1}{4}$.

Answer

$\dfrac{3}{20}$

Solution

Multiplying the numerators and denominators we have

$$\frac{3}{5} \times \frac{1}{4} = \frac{3 \times 1}{5 \times 4} = \frac{3}{20}$$

as needed.

Problem 1.16 Multiply and simplify $10 \times \dfrac{4}{5}$.

Answer

8

Solution 1

To multiply a fraction by an integer we multiply the numerator, so we have

$$10 \times \frac{4}{5} = \frac{4 \times 10}{5} = \frac{40}{5}.$$

Then noting that

$$40 \div 5 = 8$$

we have 8 as a final answer.

Solution 2

Note that 10 is a multiple of 5 with $10 = 2 \times 5$. Therefore we can cancel this 5 with the 5 in the denominator and we have

$$10 \times \frac{4}{5} = 2 \times 5 \times \frac{4}{5} = 2 \times 4 = 8$$

as our final answer.

Problem 1.17   **What is $\dfrac{5}{7} \times \dfrac{3}{2}$? Write your answer as a mixed number.**

Answer

$1\dfrac{1}{14}$

Solution

To multiply the fractions we need to multiply the numerators and denominators separately. Therefore

$$\frac{5}{7} \times \frac{3}{2} = \frac{5 \times 3}{7 \times 2} = \frac{15}{14}.$$

Since $15 \div 14$ is 1 with remainder 1, we then convert to a mixed number:

$$\frac{15}{14} = 1 + \frac{1}{14} = 1\frac{1}{14},$$

which is our answer in the correct form.

Problem 1.18   **Convert the decimal 0.02 to a simplified fraction.**

Answer

$\dfrac{1}{50}$

Solution

We first have

$$0.02 = \frac{2}{100}.$$

Since 2 is a factor of 100, we can divide the numerator and denominator by 2 to get

$$\frac{2}{100} = \frac{2 \div 2}{100 \div 2} = \frac{1}{50}$$

as our answer in simplified form.

Problem 1.19   **Convert $\dfrac{7}{20}$ into a decimal.**

Answer

0.35.

Solution 1

Note that 20 is a factor of 100, with $20 \times 5 = 100$. Therefore, multiplying the numerator and denominator by 5 we have

$$\frac{7}{20} = \frac{7 \times 5}{20 \times 5} = \frac{35}{100}.$$

Since this fraction has 100 in the denominator, it is easy to convert to a decimal:

$$\frac{35}{100} = 0.35$$

as needed.

Solution 2

Using long division we see that

$$
\begin{array}{r}
0.35 \\
20\,\overline{)\,7.00} \\
6\,0 \\
\hline
1\,00 \\
1\,00 \\
\hline
0
\end{array}
$$

so as a decimal $\frac{7}{20}$ is 0.35.

Problem 1.20 **Write the decimal 11.5 as an improper fraction.**

Answer

$$\frac{23}{2}$$

Solution 1

We start by writing 11.5 as a mixed number. We have

$$0.5 = \frac{5}{10}$$

and since 5 is a factor of 10,

$$\frac{5}{10} = \frac{5 \div 5}{10 \div 5} = \frac{1}{2}.$$

Hence

$$11.5 = 11\frac{1}{2}.$$

Since

$$11 \times 2 = 22$$

we then have

$$11\frac{1}{2} = 11 + \frac{1}{2} = \frac{22}{2} + \frac{1}{2} = \frac{23}{2}$$

as needed.

Solution 2

Recognizing that multiplying by 2 will remove the decimal, note that

$$11.5 \times 2 = 23.$$

Therefore,

$$23 \div 2 = 11.5$$

so we can write

$$11.5 = \frac{23}{2}$$

as our improper fraction.

Problem 1.21   List the common factors of 16 and 28.

Answer

1, 2, 4

Solution

The factors of 16 are
$$1, 2, 4, 8, 16,$$
and the factors of 28 are
$$1, 2, 4, 7, 14, 28.$$
Therefore the common factors are 1, 2, and 4.

Problem 1.22 What is the greatest common factor of 48 and 60?

Answer

12

Solution

The factors of 48 are
$$1, 2, 3, 4, 6, 8, 12, 16, 24, 48$$
and the factors of 60 are
$$1, 2, 3, 4, 5, 6, 10, 12, 15, 20, 30, 60.$$
The common factors of 48 and 60 are thus
$$1, 2, 3, 4, 6, 12$$
so the GCF is 12.

Problem 1.23 List the first 5 common multiples of 2 and 4.

Answer

$4, 8, 12, 16, 20$

Solution

The multiples of 2 are all even numbers. Since all multiples of 4 are even, they are all common multiples of 2 and 4. Therefore the first 5 multiples of 4,

$$4, 8, 12, 16, 20$$

are the first 5 common multiples of 2 and 4.

Problem 1.24   Find the least common multiple of 10 and 15.

Answer

30

Solution

The multiples of 10 are

$$10, 20, 30, 40, \ldots$$

and the multiples of 15 are

$$15, 30, 45, \ldots$$

so the LCM of 10 and 15 is 30.

Problem 1.25   Write the fraction $\dfrac{20}{25}$ in lowest terms.

Answer

$$\frac{4}{5}$$

Solution

To reduce a fraction, we need to find the GCF of the numerator and the denominator. The factors of 20 are

$$1, 2, 4, 5, 10, 20$$

and the factors of 25 are

$$1, 5, 25.$$

Therefore the GCF is 5. Dividing the numerator and denominator by 5 we get

$$\frac{20}{25} = \frac{20 \div 5}{25 \div 5} = \frac{4}{5}$$

as the fraction in lowest terms.

Problem 1.26 Write the fraction $\dfrac{24}{36}$ in lowest terms.

Answer

$$\frac{2}{3}$$

Solution 1

To reduce the fraction we want to find the GCF of 24 and 36. 36 is not a multiple of 24, so the GCF is less than 24. Since factors come in pairs and

$$24 = 2 \times 12,$$

the next largest factor of 24 is 12. Note that $36 = 3 \times 12$ is a multiple of 3. Therefore 12 is the GCF of of 24 and 36. Hence

$$\frac{24}{36} = \frac{24 \div 12}{36 \div 12} = \frac{2}{3}$$

is the fraction in lowest terms.

Solution 2

The factors of 24 are

$$1, 2, 3, 4, 6, 8, 12, 24,$$

and the factors of 36 are

$$1, 2, 3, 4, 6, 9, 12, 18, 36.$$

Thus the greatest common factor of 24 and 36 is 12. To simplify the fraction we divide the numerator and denominator by the GCF. Hence

$$\frac{24}{36} = \frac{24 \div 12}{36 \div 12} = \frac{2}{3}$$

is the simplified fraction.

Problem 1.27 **Write the decimal** 0.48 **as a fraction in lowest terms.**

Answer

$$\frac{12}{25}$$

Solution 1

We first have

$$0.48 = \frac{48}{100}.$$

If we recognize that 4 is a common factor of 48 and 100 we have

$$\frac{48}{100} = \frac{48 \div 4}{100 \div 4} = \frac{12}{25}.$$

Can this be simplified further? Well the factors of 12 are

$$1, 2, 3, 4, 6, 12$$

and the factors of 25 are

$$1, 5, 25.$$

Hence 1 is the only common factor, so the GCF of 12 and 25 is 1. This means that the fraction cannot be simplified further. Hence

$$0.48 = \frac{12}{25}$$

when written as a fraction in lowest terms.

Solution 2

We first have

$$0.48 = \frac{48}{100}.$$

To simplify the fraction we need to find the GCF of 48 and 100. The factors of 48 are

$$1, 2, 3, 4, 6, 8, 12, 16, 24, 48$$

and the factors of 100 are

$$1, 2, 4, 5, 10, 20, 25, 50, 100.$$

The largest number that is a factor of both, 4, is the GCF. Thus to simplify the fraction we have

$$\frac{48}{100} = \frac{48 \div 4}{100 \div 4} = \frac{12}{25}.$$

Therefore

$$0.48 = \frac{12}{25}$$

when written in lowest terms.

Problem 1.28   Convert the fraction $\dfrac{5}{8}$ to a decimal.

Answer

0.625

Solution 1

We want to write $\dfrac{5}{8}$ as a fraction with denominator 10, 100, 1000, etc. so it is easy to convert to a decimal. Note that

$$10 \div 8 = 1 \text{ remainder } 2,$$

so 8 is not a factor of 10. Similarly 8 is not a factor of 100 as

$$100 \div 8 = 12 \text{ remainder } 4.$$

We do have that

$$1000 \div 8 = 125 \text{ remainder } 0.$$

Therefore multiplying the numerator and denominator by 125 gives us

$$\frac{5}{8} = \frac{5 \times 125}{8 \times 125} = \frac{625}{1000}.$$

Thus we can write $\dfrac{5}{8}$ as the decimal 0.625.

Solution 2

Using long division,

$$
\begin{array}{r}
0.625 \\
8\overline{\smash{\big)}\,5.000} \\
\underline{4.8} \\
20 \\
\underline{16} \\
40 \\
\underline{40} \\
0
\end{array}
$$

and therefore $\dfrac{5}{8}$ is 0.625 as a decimal.

Problem 1.29 **Calculate the sum:** $\dfrac{5}{12} + \dfrac{1}{6}$.

Answer

$$\dfrac{7}{12}$$

Solution

To add the fractions, we first need to find a common denominator. This common denominator will be the LCM of 12 and 6. Note that 12 is a multiple of 6, so the least common multiple is just 12. Since $6 \times 2 = 12$, we have

$$\frac{1}{6} = \frac{1 \times 2}{6 \times 2} = \frac{2}{12}.$$

Then

$$\frac{5}{12} + \frac{2}{12} = \frac{7}{12}$$

which is our final answer.

Problem 1.30 **Calculate** $\dfrac{1}{4} - \dfrac{1}{10}$.

Answer

$$\dfrac{3}{20}.$$

Solution

To subtract the two fractions we need to find a common denominator. The multiples of 4 are

$$4, 8, 12, 16, 20, 24, \ldots$$

while the multiples of 10 are

$$10, 20, 30, \ldots$$

Therefore the least common multiple of 4 and 10 is 20. Since $4 \times 5 = 20$ we can write

$$\frac{1}{4} = \frac{1 \times 5}{4 \times 5} = \frac{5}{20}.$$

Similarly, $20 = 10 \times 2$ so

$$\frac{1}{10} = \frac{1 \times 2}{10 \times 2} = \frac{2}{20}.$$

Hence we have,

$$\frac{1}{4} - \frac{1}{10} = \frac{5}{20} - \frac{2}{20} = \frac{3}{20},$$

our final answer.

Problem 1.31 **What is the sum of $\dfrac{1}{10}$ and $\dfrac{1}{15}$? Write your answer as a simplified fraction.**

Answer

$$\frac{1}{6}$$

Solution

To add the fractions, we first find a common denominator. The multiples of 10 are

$$10, 20, 30, 40, 50, \ldots$$

and the multiples of 15 are

$$15, 30, 45, \ldots$$

---

so we see the least common multiple of 10 and 15 is 30. Since $30 = 10 \times 3$ and $30 = 15 \times 2$, we have

$$\frac{1}{10} = \frac{1 \times 3}{10 \times 3} = \frac{3}{30}$$

and

$$\frac{1}{15} = \frac{1 \times 2}{15 \times 2} = \frac{2}{30}.$$

Therefore

$$\frac{1}{10} + \frac{1}{15} = \frac{3}{30} + \frac{2}{30} = \frac{5}{30}.$$

Be careful! Note that

$$30 \div 5 = 6,$$

so 5 is a factor of 30 and the fraction can be simplified. We have

$$\frac{5}{30} = \frac{5 \div 5}{30 \div 5} = \frac{1}{6}$$

as our final answer.

Problem 1.32   What is $\dfrac{1}{3} \times \dfrac{9}{4}$? Write your answer as a simplified fraction.

Answer

$$\frac{3}{4}$$

Solution 1

Multiplying the numerators we have

$$\frac{1}{3} \times \frac{9}{4} = \frac{1 \times 9}{3 \times 4} = \frac{9}{12}.$$

To simplify, the factors of 9 are

$$1, 3, 9$$

and the factors of 12 are

$$1, 2, 3, 4, 6, 12$$

so the GCF of 9 and 12 is 3. Therefore,

$$\frac{9}{12} = \frac{9 \div 3}{12 \div 3} = \frac{3}{4},$$

which is in lowest terms.

Solution 2

If we recognize that 3 is a factor of 9, with $9 = 3 \times 3$, we can help simplify our answer immediately. We have

$$\frac{1}{3} \times \frac{9}{4} = \frac{1 \times 9}{3 \times 4} = \frac{1 \times 3 \times 3}{3 \times 4}.$$

Then, we can cancel a 3 in the numerator and denominator to get

$$\frac{3}{4},$$

which is already simplified.

Problem 1.33   **Calculate** $3\frac{1}{2} \times 2\frac{2}{3}$ **and write your answer as a simplified improper fraction.**

Answer

$\dfrac{28}{3}$

Solution

We first convert the mixed numbers into improper fractions. We have, since $3 \times 2 = 6$,

$$3\frac{1}{2} = \frac{6}{2} + \frac{1}{2} = \frac{7}{2}$$

and similarly, since $2 \times 3 = 6$,

$$2\frac{2}{3} = \frac{6}{3} + \frac{2}{3} = \frac{8}{3}.$$

Therefore, multiplying the two fractions we have

$$\frac{7}{2} \times \frac{8}{3} = \frac{7 \times 8}{2 \times 3} = \frac{56}{6}.$$

To simplify the fraction note that the factors of 6 are

$$1, 2, 3, 6.$$

Of these, only 1 and 2 are factors of 56 (as 56 is not a multiple of 3 or 6), so the GCF of 56 and 6 is 2. Hence,

$$\frac{56}{6} = \frac{56 \div 2}{6 \div 2} = \frac{28}{3},$$

written as a simplified improper fraction.

Problem 1.34 **Calculate** $\dfrac{5}{12} \div \dfrac{10}{3}$**. Express your answer as a fraction in lowest terms.**

Answer

$$\frac{1}{8}$$

Solution 1

To divide fractions we multiply by the reciprocal. Thus,

$$\frac{5}{12} \div \frac{10}{3} = \frac{5}{12} \times \frac{3}{10}.$$

We then multiply numerators and denominators,

$$\frac{5}{12} \times \frac{3}{10} = \frac{5 \times 3}{12 \times 10} = \frac{15}{120}.$$

Noting that 15 is a factors of 120, with

$$120 \div 15 = 8,$$

we can simplify our answer to

$$\frac{15}{120} = \frac{15 \div 15}{120 \div 15} = \frac{1}{8},$$

our final answer.

Solution 2

To divide fractions we multiply by the reciprocal. Thus,

$$\frac{5}{12} \div \frac{10}{3} = \frac{5}{12} \times \frac{3}{10}.$$

Note now that 5 is a factor of 10, with $10 = 5 \times 2$ and 3 is a factor of 12 with $12 = 3 \times 4$, therefore we can cancel the 5 and 3 in the numerator to get

$$\frac{5}{12} \times \frac{3}{10} = \frac{5 \times 3}{12 \times 10} = \frac{5 \times 3}{3 \times 4 \times 5 \times 2} = \frac{1}{4 \times 2} = \frac{1}{8}.$$

Since this is already simplified, it is our final answer.

Problem 1.35   **Divide** $3\frac{3}{8}$ **by 3. Write your answer as a mixed number.**

Answer

$1\frac{1}{8}$

Solution 1

We first convert $3\frac{3}{8}$ to a improper fraction so we can divide:

$$3\frac{3}{8} = 3 + \frac{3}{8} = \frac{3 \times 8}{1 \times 8} + \frac{3}{8} = \frac{24}{8} + \frac{3}{8} = \frac{27}{8}.$$

Dividing by 3 is the same as multiplying by $\frac{1}{3}$, so

$$\frac{27}{8} \div 3 = \frac{27}{8} \times \frac{1}{3}.$$

Multiplying the numerator and denominator we have

$$\frac{27}{8} \times \frac{1}{3} = \frac{27 \times 1}{8 \times 3} = \frac{27}{24}.$$

Since

$$27 \div 24 = 1 \text{ remainder } 3,$$

we have

$$\frac{27}{24} = 1\frac{3}{24}.$$

Lastly, 3 is a factor of $24 = 3 \times 8$, so

$$1\frac{3}{24} = 1\frac{3 \div 3}{24 \div 3} = 1\frac{1}{8}$$

is our final answer.

Solution 2

Recall that

$$3\frac{3}{8} = 3 + \frac{3}{8}.$$

Therefore distributing we have

$$3\frac{3}{8} \div 3 = (3 + \frac{3}{8}) \div 3 = 3 \div 3 + \frac{3}{8} \div 3.$$

Clearly,

$$3 \div 3 = 1$$

so we are left to simplify $\frac{3}{8} \div 3$. Dividing is the same as multiplying by the reciprocal, so

$$\frac{3}{8} \div 3 = \frac{3}{8} \times \frac{1}{3} = \frac{1}{8}$$

as the 3's cancel. Hence

$$3\frac{3}{8} \div 3 = 1 + \frac{1}{8} = 1\frac{1}{8}$$

written as a mixed number.

Problem 1.36 **Mr. and Mrs. Smith are both teachers. Mr. Smith's class has** 24 **students, while Mrs. Smith's class has** 30 **students. Mr. Smith divides his class into** N **groups, with each group having the same size. Mrs. Smith also divides her class into** N **groups, with each group having the same size. What is the largest value of** N **so this is true?**

Answer

6

Solution

If Mr. Smith divides his 24 students into N groups of the same size, then N must be a factor of 24. Similarly, for Mrs. Smith, we see N must be a factor of 30. Since we want the largest possible N, we need to find the GCF of 24 and 30. The factors of 24 are

$$1, 2, 3, 4, 6, 8, 12, 24,$$

and the factors of 30 are

$$1, 2, 3, 5, 6, 10, 15, 30.$$

Thus the greatest common factor of 24 and 30 is 6. Hence, $N = 6$ is the largest so that both Mr. and Mrs. Smith can divide their classes into N groups.

Problem 1.37 Billy has 60 green marbles and 40 blue marbles. He wants to put the marbles in bags so that each bag as the same number of green and blue marbles. Billy wants to use as many bags as possible, but wants to make sure that each bag has the same number of green and blue marbles. How many bags should Billy use?

Answer

20

Solution

For each bag to have the same number of green marbles, the number of bags must be a factor of 60. Similarly, for each bag to have the same number of blue marbles, the number of bags must be a factor of 40. Hence we want to find the GCF of 60 and 40. Recall that factors come in pairs, so the largest factors of 60 are

$$60, 30, 20, \ldots,$$

because

$$60 = 1 \times 60 = 2 \times 30 = 3 \times 20.$$

60 and 30 are not factors of 40, but

$$40 \div 20 = 2,$$

so 20 is a factor of 40 and hence the GCF of 60 and 40. Therefore Billy should use 20 bags in total.

Problem 1.38 Nick and Justin run laps around the school. Nick completes a lap every 4 minutes while Justin completes a lap every 6 minutes. They agree to run until they meet up for a second time at the starting line. How many minutes in total do they run?

Answer

24

Solution

Nick completes a lap every 4 minutes, so he will be at the starting line after

$$4, 8, 12, 16, 20, 24, 28, 32, 36, 40, \ldots$$

minutes, every multiple of 4. Similarly, Justin will be at the starting line after

$$6, 12, 18, 24, 30, 36, \ldots$$

minutes, every multiple of 6. They will both be at the starting line at common multiples of 4 and 6, which are

$$12, 24, 36, \ldots$$

The second smallest common multiple is 24, so Nick and Justin stop running after 24 minutes.

Problem 1.39 **Grace and Andrew enjoy building towers with blocks. Grace has blocks that are 10 cm high and Andrew has blocks that are 6 cm high. What is the smallest height that both Grace and Andrew can build with they blocks?**

Answer

30 cm

Solution

Grace has blocks that are 10 cm high, so she can build towers that have a height that is a multiple of 10. Similarly, Andrew can build towers that have a height that is a multiple of 6. Hence the smallest height they can both build will be the LCM of 10 and 6. Multiples of 10 are

$$10, 20, 30, 40, \ldots$$

while multiples of 6 are

$$6, 12, 18, 24, 30, 36, \ldots$$

Therefore, the LCM of 10 and 6 is 30 and 30 cm is the smallest height they Grace and Andrew can both build.

Problem 1.40 **Jane's family loves to come visit her. Her parent's come visit every 6 weeks, and her grandparents come visit every 8 weeks. How often do her parents and grandparents both visit the same week?**

Answer

Every 24 weeks.

Solution

Jane's parents visit every 6 weeks and grandparents every 8 weeks. They both will visit on weeks that are a common multiple of 6 and 8. Thus they will visit every week that is a multiple of the LCM of 6 and 8. Multiples of 6 are

$$6, 12, 18, 24, 30, 36, \ldots$$

and multiples of 8 are

$$8, 16, 24, 32, \ldots$$

so the LCM of 6 and 8 is 24. Thus Jane's parents and grandparents visit together every 24 weeks.

Problem 1.41 **Find the largest whole number less than 40 that leaves a remainder of 2 when divided by 7.**

Answer

37

Solution 1

Looking at multiples of 7 less than 40, we have the numbers

$$7, 14, 21, 28, 35.$$

All of these number leave 0 remainder when divided by 7. Hence the numbers

$$7 + 2 = 9, 14 + 2 = 16, 21 + 2 = 23, 28 + 2 = 30, 35 + 2 = 37$$

all have remainder 2 when divided by 7. The largest less than 40 is 37.

Solution 2

Consider the number 40, we have

$$40 \div 7 = 5 \text{ remainder } 5,$$

so 40 leaves a remainder of 5 when divided by 7. We want a remainder of 2, which is

$$5 - 2 = 3$$

less than 5. Hence the number

$$40 - 3 = 37$$

will leave the correct remainder of 2 when divided by 7.

Problem 1.42 **Find the greatest 2-digit number that leaves a remainder of** 4 **when divided by** 11.

Answer

92

Solution 1

Looking at the 2-digit multiples of 11, they are

$$11, 22, 33, 44, 55, 66, 77, 88, 99.$$

All of these numbers leave remainder of 0 when divided by 11. Adding 4 to each of these, the numbers

$$15, 26, 37, 48, 59, 70, 81, 92, 103$$

all leave remainder of 4 when divided by 11. Note that while 99 is less than 100, 103 is not. Hence the largest 2-digit number that leaves a remainder of 4 when divided by 11 is 92.

Solution 2

Consider the number 100. We have

$$100 \div 11 = 9 \text{ remainder } 1.$$

This means decreasing 100 by 1 to 99 will give a number that has remainder 0 when divided by 11. Decreasing 99 to 98, we have

$$98 \div 11 = 8 \text{ remainder } 10$$

because 98 is one less than a multiple of 11 and

$$11 - 1 = 10.$$

Since we want a number with remainder 4, and

$$10 - 4 = 6,$$

the number

$$98 - 6 = 92$$

is the largest 2-digit number that has remainder 4 when divided by 11.

Problem 1.43 **Find the smallest whole number that leaves a remainder of** 2 **when divided by** 5 **and a remainder of** 4 **when divided by** 6.

Answer

22

Solution

Multiples of 5 leave no remainder when divided by 5, so adding 2 to a multiple of 5 will give a number that has remainder 2 when divided by 5. Hence,

$$5 + 2 = 7, 10 + 2 = 12, 15 + 2 = 17, 20 + 2 = 22, 25 + 2 = 27, \dots$$

all have remainder 2 when divided by 5. Similarly, the numbers

$$6 + 4 = 10, 12 + 4 = 16, 18 + 4 = 22, 24 + 4 = 28, \dots$$

all have remainder 4 when divided by 6. Comparing these lists, we see that 22 is the smallest number that leaves remainder 2 when divided by 5 and 4 when divided by 6.

Problem 1.44 **How many of the improper fractions**

$$\frac{20}{1}, \frac{20}{2}, \frac{20}{3}, \frac{20}{4}, \ldots, \frac{20}{20}$$

can be simplified to whole numbers?

Answer

6

Solution

An improper fraction can be simplified to a whole number when the denominator is a factor of the numerator. The numerators are all 20, which has factors

$$1, 2, 4, 5, 10, 20.$$

Hence only the fractions

$$\frac{20}{1} = 20, \frac{20}{2} = 10, \frac{20}{4} = 5, \frac{20}{5} = 4, \frac{20}{10} = 2, \frac{20}{20} = 1,$$

a total of 6, can be simplified to whole numbers.

Problem 1.45 **Josh goes shopping for socks. Each pair of socks costs $8. Josh buys as many socks as he can with $100. How many pairs of socks does Josh buy? What is his change?**

Answer

12 pairs of socks, $4 change

Solution

Josh has a total of 100 dollars. Since each pair of socks is $8, we look at 100 divided by 8, which is

$$100 \div 8 = 12 \text{ remainder } 4,$$

because
$$12 \times 8 = 96.$$

Therefor Josh can buy 12 pairs of socks for $96 and the remaining

$$100 - 96 = 4$$

dollars is change.

Problem 1.46 **Emily Elizabeth is painting rainbow eggs for Easter. She painted Red, Orange, Yellow, Green, Gray, Blue, and Purple colors in order for a palette of** 80 **eggs delivered by Farmer Brown. What is the color of the last egg?**

Answer

Yellow

Solution

There are 7 colors in total. Since Emily paints the eggs in order, every 7th egg will be purple. There are 80 eggs in total, and since

$$80 \div 7 = 11 \text{ remainder } 3$$

egg number

$$11 \times 7 = 77$$

will also be purple. There are then 3 eggs remaining, so the last egg will be the third color, yellow.

Problem 1.47 **Four friends, Alice, Bob, Charlie, and Drew, count to** 50, **as in the chart below:**

A	B	C	D
1	2	3	4
8	7	6	5
9	10	11	12
...	14	13

Which of the four friends says the number 50?

Answer

Bob

Solution

There are 4 friends in total, but because the counting goes $A, B, C, D, D, C, B, A, \ldots$ the pattern repeats every 8 numbers. Therefore Alice is the one who counts of

$$8, 16, \ldots$$

Note that

$$50 \div 8 = 6 \text{ remainder } 2$$

so the pattern repeats 6 times by the time Alice says

$$8 \times 6 = 48.$$

Two more number remain, so the second friend Bob is the one who says 50.

Problem 1.48 **George wrote a computer program. If you type in a positive integer, George's program outputs the sum of all the positive factors of that integer. For example, if you type in 6, the program outputs 12, because the factors of 6 are $1, 2, 3, 6$ and $1 + 2 + 3 + 6 = 12$. Suppose George's friend Paul types in 7 and then runs the program. If Paul types in this result and runs the program again, what is the output?**

Answer

15

Solution

The number 7 is a prime number, so the only factors of 7 are 1 and 7. Therefore, when Paul runs the program the first time, the output is

$$1 + 7 = 8.$$

8 has factors

$$1, 2, 4, 8$$

so when he runs it a second time the output is

$$1 + 2 + 4 + 8 = 15.$$

Hence the answer is 15.

Problem 1.49 Consider the number 15. Multiply together all the factors of 15. What is the result?

Answer

225

Solution 1

Recall that factors of 15 come in pairs. We have

$$15 = 1 \times 15 = 3 \times 5$$

so the factors are $1, 15, 3, 5$. This makes it easy to multiply as

$$1 \times 15 \times 3 \times 5 = (1 \times 15) \times (3 \times 5) = 15 \times 15 = 225.$$

Solution 2

Listing the factors of 15 we see they are

$$1, 3, 5, 15.$$

We can then multiply them out to get

$$1 \times 3 \times 5 \times 15 = 225$$

as their product.

Problem 1.50 Frank's goal is the drink 5 liters of water per day. He drinks water out of a bottle that holds $\frac{2}{3}$ of a liter of water. How many bottles of water does Frank need to drink to reach his goal of 5 liters? Express your answer as a mixed number.

Answer

$7\frac{1}{2}$

Solution

Each water bottle contains $\dfrac{2}{3}$ of a liter, so we need to divide to find out how many water bottles Frank should drink. Recalling that dividing by a fraction is the same as multiplying by the reciprocal,

$$5 \div \frac{2}{3} = 5 \times \frac{3}{2} = \frac{15}{2}.$$

Converting this improper fraction to a mixed number, note

$$15 \div 2 = 7 \text{ remainder } 1,$$

so Frank needs to drink

$$\frac{15}{2} = 7\frac{1}{2}$$

bottles of water to reach his goal.

Problem 1.51 **Julie buys lemons and limes at the grocery store. The price of lemons is 4 lemons per $1 and the price of limes is 6 limes for $1. Julie buys a total of 6 lemons and 21 limes. How much money does Julie spend in total?**

Answer

$5

Solution

Lemons are 4 for 1 dollar, so each lemon is

$$1 \div 4 = \frac{1}{4}$$

of a dollar. Similarly, each lime is

$$1 \div 6 = \frac{1}{6}$$

of a dollar. Julie buys 6 lemons, so Julie spends a total of

$$6 \times \frac{1}{4} = \frac{6}{4}$$

dollars on lemons. Note that 6 and 4 are even, so we can divide by two to simplify,

$$\frac{6}{4} = \frac{6 \div 2}{4 \div 2} = \frac{3}{2}.$$

She also buys 21 limes, so the amount of money she spends on limes is

$$21 \times \frac{1}{6} = \frac{21}{6}$$

dollars. To simplify this fraction, note 3 is the GCF of 21 and 6. Therefore,

$$\frac{21}{6} = \frac{21 \div 3}{6 \div 3} = \frac{7}{2}.$$

In total, Julie spends

$$\frac{3}{2} + \frac{7}{2} = \frac{10}{2} = 5$$

dollars at the grocery store.

Problem 1.52 **Tom is planting a tree and needs to buy soil. He knows he needs $3\frac{2}{3}$ cubic feet of soil and the soil costs \$3.50 dollars per cubic foot. How much money will Tom need to spend on the soil? Express your answer as a mixed number.**

Answer

$12\frac{5}{6}$

Solution

It is easiest to multiply fractions as improper fractions, so we first convert both numbers to improper fractions. Since

$$3 \times 3 = 9,$$

we have

$$3\frac{2}{3} = 3 + \frac{2}{3} = \frac{9}{3} + \frac{2}{3} = \frac{11}{3}$$

for the number of cubic feet of soil Tom needs for his tree. Similarly,

$$3.50 = 3\frac{1}{2} = 3 + \frac{1}{2} = \frac{6}{2} + \frac{1}{2} = \frac{7}{2}.$$

Hence the total amount Tom needs to spend is

$$\frac{11}{3} \times \frac{7}{2} = \frac{11 \times 7}{3 \times 2} = \frac{77}{6}.$$

Since

$$77 \div 6 = 12 \text{ remainder } 5,$$

as a mixed number Tom must spend

$$\frac{77}{6} = 12\frac{5}{6}$$

dollars on new soil to plant his tree.

Problem 1.53 **Florence has some candy bars. If she keeps 3 for herself, she can distribute the remaining candy bars evenly among 4 children. If she keeps 5 for herself, she can distribute the remaining candy bars evenly among 9 children. If Florence has less than 30 candy bars, exactly how many does she have?**

Answer

23

Solution

We know Florence can distribute the candy bars evenly among 4 people if she keeps 3 for herself, so the number of candy bars has remainder 3 when divided by 4. The number of candy bars is thus 3 more than a multiple of 4. The numbers less than 30 of this form are

$$4+3 = 7, 8+3 = 11, 12+3 = 15, 16+3 = 19, 20+3 = 23, 24+3 = 27.$$

Distributing among 9 children, we know she can keep 5 for herself, so the number of candy bars leaves a remainder of 5 when divided by 9. Similar to above we can then list the possibilies less than 30:

$$9+5 = 14, 18+5 = 23.$$

Note that 23 is the only number in both lists, so we know that Florence must have exactly 23 candy bars.

Problem 1.54   Convert the fraction $\dfrac{2}{9}$ **to a decimal.**

$0.\overline{2}$

Using long division we have

$$
\begin{array}{r}
0.22 \\
9\overline{\smash{\big)}\,2.00} \\
\underline{1.8} \\
20 \\
\underline{18} \\
2
\end{array}
$$

From here, note that the pattern of dropping the 0 to get 20, noting that $18 = 9 \times 2$ is the smallest multiple of 9 less than 20, and subtracting to get

$$20 - 18 = 2,$$

will repeat over and over. Hence

$$\frac{2}{9} = 0.2222\cdots = 0.\overline{2}$$

as needed.

Problem 1.55   Express the fraction $\dfrac{5}{11}$ **as a repeating decimal.**

$0.\overline{45}$

Using long division we have

$$
\begin{array}{r}
0.4545 \\
11 \overline{\smash{\big)}\ 5.0000} \\
4.4 \\
\hline
60 \\
55 \\
\hline
50 \\
44 \\
\hline
60 \\
55 \\
\hline
5
\end{array}
$$

Notice that the pattern of remainders of $6, 5, 6, 5$ will keep continuing, leading to a decimal that repeats every 2 digits. We therefore have

$$
\frac{5}{11} = 0.454545\cdots = 0.\overline{45}
$$

as our repeating decimal.

Problem 1.56 **Write the fraction $\dfrac{5}{6}$ as a decimal.**

Answer

$0.8\overline{3}$

Solution

Using long division we have

$$
\begin{array}{r}
0.833 \\
6 \overline{\smash{\big)}\ 5.000} \\
4.8 \\
\hline
20 \\
18 \\
\hline
20 \\
18 \\
\hline
2
\end{array}
$$

There is initially no pattern, but from the hundreds digit onwards we have repeated remainders of 2, leading to repeated digits 3. Hence we have

$$\frac{5}{6} = 0.83333\cdots = 0.8\overline{3}$$

written as a repeating decimal.

Problem 1.57 **Convert the decimal $0.\overline{6}$ into a fraction.**

Answer

$$\frac{2}{3}.$$

Solution

Let

$$x = 0.66666\cdots = 0.\overline{6}.$$

Note that

$$10x = 6.6666\cdots = 6.\overline{6},$$

so that

$$10x - x = 6.6666\cdots - 0.66666\cdots = 6.\overline{6} - 0.\overline{6} = 6.$$

Simplifying the left side we get

$$9x = 6$$

so

$$x = \frac{6}{9}.$$

Note that 3 is a common factor of $6, 9$ so

$$x = \frac{6}{9} = \frac{6 \div 3}{9 \div 3} = \frac{2}{3}.$$

Hence $0.\overline{6} = \frac{2}{3}$.

Problem 1.58 **Write the repeating decimal $0.2\overline{6}$ as a simplified fraction.**

Answer

$$\frac{4}{15}$$

Solution

Let

$$x = 0.266666\cdots = 0.2\overline{6}.$$

Multiplying by 10 we get

$$10x = 2.66666\cdots = 2.\overline{6}$$

so the decimal portion is repeating. Multiplying by 10 again we have

$$100x = 26.6666\cdots = 26.\overline{6}.$$

Now note that

$$100x - 10x = 26.6666\cdots - 2.66666\cdots = 26.\overline{6} - 2.\overline{6} = 24.$$

Simplifying we have

$$90x = 24,$$

so

$$x = \frac{24}{90}.$$

The GCF of 24 and 90 is 6, so simplifying we have

$$\frac{24}{90} = \frac{24 \div 6}{90 \div 6} = \frac{4}{15},$$

so

$$0.2\overline{6} = \frac{4}{15}$$

written as a fraction.

Problem 1.59 **Write the repeating decimal $0.\overline{21}$ as a fraction.**

Answer

$$\frac{7}{33}$$

Solution

Let
$$x = 0.2121212121\cdots = 0.\overline{21}.$$

Multiplying by 100 we have
$$100x = 21.21212121\cdots = 21.\overline{21}$$

so we have the same repeated decimal portion. Therefore,
$$100x - x = 21.21212121\cdots - 0.2121212121\cdots = 21.\overline{21} - 0.\overline{21} = 21.$$

Combining like terms,
$$99x = 21$$

so
$$x = \frac{21}{99}.$$

The factors of 21 are
$$1, 3, 7, 21.$$

Of these, 3 is the largest that is also a factor of 99. Hence
$$x = 0.\overline{21} = \frac{21}{99} = \frac{21 \div 3}{99 \div 3} = \frac{7}{33}$$

written as a fraction in lowest terms.

Problem 1.60 **Write the number $2.\overline{2}$ as a mixed number.**

Answer

$2\dfrac{2}{9}$

Solution

Let
$$x = 2.2222\cdots = 2.\overline{2}.$$

Multiplying by 10 we have
$$10x = 22.222\cdots = 22.\overline{2}.$$

Thus
$$10x - x = 22.222 \cdots - 2.2222 \cdots = 22.\overline{2} - 2.\overline{2} = 20.$$

Thus
$$9x = 20,$$

so
$$x = \frac{20}{9}.$$

Since
$$20 \div 9 = 2 \text{ remainder } 2,$$

as a mixed number,
$$2.\overline{2} = 2\frac{2}{9}.$$

2. Ratio, Proportion, and Percentage

Problem 2.1 **Last month the price of gas was $1.10 per gallon. This month gas is selling for $1.32 per gallon. Find the percentage increase in the price of gas per gallon.**

Answer

20%

Solution

First we find the difference between the two prices, which is

$$1.32 - 1.1 = 0.22$$

dollars. We then divide by the original price to get the percentage increase

$$0.22 \div 1.10 = 0.2 = 20\%.$$

Therefore, the percentage increase in the price of gas is 20%.

Problem 2.2 **There are** 180 **days in a school year. John was present** 85% **of the total days. How many days was he present?**

Answer

153

Solution

John was present 85% of the 180 days, so multiplying we get

$$180 \times 85\% = 180 \times 0.85 = 153,$$

so John was present 153 days.

Problem 2.3 **Sam received a grade of** 80% **on a geometry test. If he solved** 24 **problems correctly, how many problems were on the test?**

Answer

30

Solution 1

Since we know that 80% of the total number of problems is 24, we divide to get the test had

$$24 \div 80\% = 24 \div 0.80 = 30$$

problems in total

Solution 2

(Algebra) Let x be total number of problems on the test. Then

$$80\% \times x = 24$$

Solving the equation,

$$x = \frac{24}{0.80} = 30.$$

Thus, there we were 30 problems on the test.

Problem 2.4 **Jim receives a weekly salary of** $200. **He spends** $60 **per week on gas. What percent of his weekly salary does he use for gas?**

Answer

30%

Solution

Use division to find the percentage,

$$60 \div 200 = 0.3 = 30\%,$$

so Jim uses 30% of his weekly salary on gas.

Problem 2.5 **Last year Areteem Institute had an enrollment of** 500 **students. This year the enrollment is** 800 **students. What is the percent of increase in student enrollment?**

Answer

60%

Solution

The difference in enrollments between the two years is

$$800 - 500 = 300$$

students. Dividing by the original (last year's) enrollment

$$300 \div 500 = 0.6 = 60\%.$$

Hence, the percent of increase in student enrollment is 60%.

Problem 2.6 **Jess paid** $6.31 **for a shirt marked** 25% **off the regular price. What was the regular price of the shirt?**

Answer

$11.20

Solution 1

After 25% off, Jess paid $8.40, so $8.40 is

$$100\% - 25\% = 75\%$$

of the regular price. Therefore we divide to get

$$8.40 \div 0.75 = 11.2$$

so the regular price of the shirt is $11.20.

Solution 2

Let the regular price to be x. Then,

$$(1 - 0.25) \times x = 8.4.$$

Solve for x, then

$$x = \frac{8.4}{0.75} = 11.2.$$

So the regular price of the shirt is $11.2.

Problem 2.7   **John bought a coat which usually sells for $98.00 at 25% off. What did he pay for the coat?**

Answer

$73.50

Solution 1

The discount is
$$98 \times 25\% = 98 \times 0.25 = 24.5$$

dollars. Therefore John paid Use the original price minus 25% of the original price,
$$98 - 24.5 = 73.5$$

dollars for the coat.

Solution 2

25% off the price means we need to pay

$$100\% - 25\% = 75\%$$

of the original price. Therefore the discounted price is

$$98 \times 0.75 = 73.5.$$

so John paid $73.50 for the coat.

Problem 2.8 **A coffee mug marked $15.00 was on sale for $12.00. What is the rate of discount?**

Answer

20%

Solution 1

The discount of the coffee mug was

$$15.00 - 12.00 = 3.00$$

dollars. Since the original price was $15.00, the rate of discount is

$$3.00 \div 15.00 = 0.2 = 20\%,$$

so the rate of discount is 20%.

Solution 2

The sale price was

$$12.00 \div 15.00 = 0.8 = 80\%$$

of the original price. Therefore the rate of discount is

$$100\% - 80\% = 20\%.$$

Problem 2.9 **A salesman who works on a commission basis earns** 18% **of his sales. How much was his commission on a** $480 **sale?**

Answer

$86.40

Solution

18% of 480 is
$$480 \times 18\% = 480 \times 0.18 = 86.4.$$
Therefore, the salesman's commission was $86.40.

Problem 2.10 **A clerk at the stationary store receives** $12\frac{1}{2}\%$ **commission on all merchandise sold. If she received** $52 **in commission last week, what were her sales for the week?**

Answer

$416

Solution 1

The clerk earned $52 dollars. Since this is $12\frac{1}{2}\%$ of the total sales, we divide to get
$$52 \div 12\frac{1}{2}\% = 52 \div 0.125 = 416.$$
Hence, her sales for last week were $416.

Solution 2

Let her sales to be x dollars, then
$$12\frac{1}{2}\% \times x = 52.$$
Solve for x to get
$$x = \frac{52}{0.125} = 416,$$
so her sales were $416.

Problem 2.11   **When gold sold for $16 an ounce, Johnny found $6 worth of gold in his claim. Gold presently sells for $328 an ounce. How many dollars is Johnny's amount of gold worth today?**

Answer

$123

Solution 1

Dividing the new price per ounce by the original price per ounce we have that the new price is

$$328 \div 16 = 20.5 = 2050\%$$

of the old price. Therefore Johnny's $6 of gold is now worth

$$2050\% \times 6 = 20.5 \times 6 = 123$$

dollars.

Solution 2

Johnny found gold originally worth $6. Since gold was $16 an ounce, this means Johnny found

$$6 \div 16 = \frac{3}{8} = 0.375$$

of an ounce of gold. Since gold is now worth $328 an ounce, Johnny's gold is worth

$$0.375 \times 328 = 123$$

dollars.

Solution 3

Let x be the dollars amount that Johnny's gold is now worth. Since the amount of gold has not changed, the ratio of the price of Johnny's gold to the price per ounce is the same. Therefore, comparing the old prices to the new prices we have

$$\frac{6}{16} = \frac{x}{328}.$$

Solve for x, then,

$$x = \frac{6}{16} \times 328 = 123.$$

Thus, Johnny's amount of gold is now worth $123.

Problem 2.12 **The ratio of girls to boys participating in intramural volleyball at Ashland Middle School is 7 to 4. There are 42 girls in the program. What is the total number of participants?**

Answer

66

Solution 1

Since the ratio of girls to boys is 7 to 4, there are 4 boys for every 7 girls and hence 7 out of every 11 participants are girls. Since $\frac{7}{11}$ of the participants are girls and there are 42 girls in total, we divide

$$42 \div \frac{7}{11} = 66.$$

so the total number of participants is 66.

Solution 2

The ratio of girls to boys is $7 : 4$. There are 42 girls, and

$$42 \div 7 = 6$$

so we can multiply the ratio by 6 on both sides to get

$$7 : 4 = 42 : 24.$$

Thus, there are 42 girls and 24 boys, for a total of

$$42 + 24 = 66$$

participants.

Solution 3

Let x be the number of boys in the program. According to the ratio of girls to boys we have

$$\frac{7}{4} = \frac{42}{x}.$$

Cross multiplying we have

$$7x = 168$$

so we can solve for x to get

$$x = \frac{168}{7} = 24.$$

Since there are 24 boys (and 42 girls) there are

$$24 + 42 = 66$$

participants in total.

Problem 2.13 **If 5 bananas are worth the same as 3 apples then how many bananas are worth the same as 15 apples?**

Answer

25

Solution 1

We are given that bananas to apples in a $5 : 3$ ratio are worth the same. Since

$$15 \div 3 = 5$$

we multiply both sides of the ratio by 5 to get

$$5 : 3 = 25 : 15$$

so 25 bananas are worth the same as 15 apples.

Solution 2

Let the number of bananas to be x. We want the ratio $x : 15$ to be equal to $5 : 3$ or

$$\frac{5}{3} = \frac{x}{15}.$$

We can then solve for x so

$$x = 15 \times \frac{5}{3} = 25.$$

Thus, 25 bananas are worth 15 apples.

Problem 2.14 **An agent receives a commission of 6% of the selling price of a house. The rest of the proceeds go to the owner of the house. If the agent sells a house for \$135,000, what is her commission? How much does the house owner receive?**

Answer

Agent: \$8100, Owner: \$126900

Solution

The agent receives a 6% commission, so she receives

$$6\% \times 135000 = 0.06 \times 135000 = 8100$$

dollars. Therefore, the house owner receives the rest, a total of

$$135000 - 8100 = 126900$$

dollars.

Problem 2.15 **The ratio of the length of Mary's cat to the length of Amy's cat is 5:7. Mary's cat is 100 cm long. How much longer is Amy's cat than Mary's cat?**

Answer

40 cm

Solution 1

The ratio of lengths is $5 : 7$. Since we know the length of Mary's cat and

$$100 \div 5 = 20$$

we multiply both sides of the ratio by 20:

$$5 : 7 = 100 : 140.$$

Therefore Amy's cat is 140 cm long, which is

$$140 - 100 = 40$$

cm longer than Mary's cat.

Solution 2

The ratio of cat lengths is $5 : 7$. Therefore, Amy's cat is

$$\frac{7}{5} = 1\frac{2}{5}$$

the length of Mary's cat. Hence, Amy's cat is $\frac{2}{5}$ longer than Amy's cat. Since

$$\frac{2}{5} \times 100 = 40,$$

Amy's cat is 40 cm longer than Mary's.

Solution 3

(Algebra) Let the length of Amy's cat to be x cm. We know the ratio of cat lengths is $5 : 7$. Since Mary's cat is 100 cm long we have

$$\frac{5}{7} = \frac{100}{x}.$$

Cross multiplying we get

$$5x = 700$$

so solving for x we have

$$x = 140.$$

The length of Amy's cat is 140 cm, which is

$$140 - 100 = 40$$

cm longer than Mary's cat.

Problem 2.16 **Thomas mixed 3 pints of red paint with 4 pints of blue paint to make a new color. He now wants to use 27 pints of red paint and some blue paint to make the same color. How many pints of blue paint will he need?**

Answer

36

Solution 1

We know the ratio of red to blue paint is $3 : 4$. Since

$$27 \div 3 = 9$$

we can multiply both sides of the ratio by 9 to get

$$3 : 4 = 27 : 36.$$

Therefore, if Thomas uses 27 pints of red paint he will need 36 pints of blue paint.

Solution 2

Let the amount of blue paint Thomas needs be x pints. Since the ratio of red to blue paint is $3 : 4$ and Thomas wants to use 27 pints of red paint we have

$$\frac{3}{4} = \frac{27}{x}.$$

Cross multiplying we get

$$3x = 108$$

so solving for x we have

$$x = 36.$$

so Thomas will need 36 pints of blue paint.

Problem 2.17 **The ratio of boys to girls at the baseball game is** $5 : 2$**. There are 28 girls. How many more boys are there than girls?**

Answer

42

Solution 1

The ratio of boys to girls is $5 : 2$. Since there are 28 girls and

$$28 \div 2 = 14$$

we can multiply both sides of the ratio by 14 to get

$$5 : 2 = 70 : 28$$

so there are 70 boys at the baseball game. Finding the difference,

$$70 - 28 = 42$$

we see there are 42 more boys than girls.

Solution 2

Since we know the ratio of boys to girls is $5 : 2$ for every 2 girls there are 3 extra boys. Since there are 28 girls in total and

$$28 \div 2 = 14$$

there are

$$14 \times 3 = 42$$

more boys than girls.

Solution 3

Since the ratio of boys to girls is $5 : 2$, 2 out of every 7 people are girls. Since $\frac{2}{7}$ of the people attending are girls, there must be

$$28 \div \frac{2}{7} = 98$$

total people in attendance. Of these,

$$98 - 28 = 70$$

are boys, so there are

$$70 - 28 = 42$$

more boys than girls.

Solution 4

(Algebra) Let the number of boys to be x. Since the ratio of boys to girls is $5 : 2$ and there are 28 girls we have

$$\frac{5}{2} = \frac{x}{28}.$$

Solving for x we have

$$x = 28 \times \frac{5}{2} = 70,$$

so there are 70 boys. Then the difference between boys and girls is

$$70 - 28 = 42,$$

so there are 42 more boys than girls.

Problem 2.18 **Cathy and Jimmy looked for seashells at Newport Beach. For every 9 seashells Cathy found, Jimmy found 7. Cathy found 54 seashells. How many fewer seashells did Jimmy find than Cathy?**

Answer

12

Solution 1

We are told the ratio of seashells found is $9 : 7$. Since Cathy found 54 seashells we calculate

$$54 \div 9 = 6.$$

Multiplying both sides of the ratio by 6 we get

$$9 : 7 = 54 : 42,$$

so Jimmy found 42 seashells. Now we can find the difference,

$$54 - 42 = 12.$$

Hence, Jimmy found 12 fewer seashells than Cathy.

Solution 2

Since the ratio of seashells found by Cathy and Jimmy is $9 : 7$, for every 9 seashells Cathy finds, Jimmy finds 2 less. Since

$$54 \div 9 = 6$$

we know that Jimmy found

$$6 \times 2 = 12$$

fewer seashells than Cathy.

Solution 3

Since the ratio of seasheels found by Cathy and Jimmy is $9 : 7$, Cathy finds 9 out of every 16 seashells. Since Cathy finds $\dfrac{9}{16}$ of the seashells and we know she finds 54 herself, there must have been

$$54 \div \frac{9}{16} = 96$$

seashells in total. Therefore Jimmy finds

$$96 - 54 = 42$$

seashells, which is

$$54 - 42 = 12.$$

fewer than Cathy.

Solution 4

Let the number of seashells that Jimmy found to be x. The ratio of seashells is $9 : 7$ so since Cathy found 54 seashells we have

$$\frac{9}{7} = \frac{54}{x}.$$

Cross multipy to get

$$9x = 378$$

so we solve for x,

$$x = 42.$$

Therefore, Jimmy finds 42 seashells which is

$$54 - 42 = 12.$$

fewer seashells than Cathy.

Problem 2.19 **There are 2 packs of crayons available for every 5 students at Amy's art class. How many students can share 18 packs of crayons?**

Answer

25

Solution 1

The ratio of packs of crayons to students is $2 : 5$. Since there are 18 packs of crayons available, if we multiply both sides of the ratio by

$$18 \div 2 = 9$$

we have

$$2 : 5 = 18 : 45$$

So, 45 students can share 18 packs of crayons.

Solution 2

Let the number of students to be x. Since the ratio of packs of crayons to students is $2 : 5$ and there are 18 packs of crayons,

$$\frac{2}{5} = \frac{18}{x}.$$

If we cross multiply we have

$$2x = 80$$

so we get

$$x = 45.$$

Hence, 45 students can share 18 packs of crayons.

Problem 2.20 **The ratio of boys to girls in Math Zoom Academy is 5:3. If there are 24 more boys than girls in Math Zoom Academy, then how many girls are in Math Zoom Academy?**

Answer

36

Solution 1

The ratio of boys to girls is $5 : 3$, so there are 2 more boys than girls for every 3 girls. Since

$$24 \div 2 = 12$$

multiplying both sides of the ratio by 12 we get

$$5 : 3 = 60 : 36$$

and there are 36 girls in Math Zoom Academy.

Solution 2

Let the number of girls in Math Zoom Academy to be x. Then number of boys in Math Zoom Academy will be $x + 24$. Since the ratio of boys to girls is $5 : 3$ we have the following equation:

$$\frac{5}{3} = \frac{x+24}{x}.$$

Cross multiplying gives us

$$5x = 3(x + 24).$$

Distributing and combining like terms we have

$$2x = 72$$

so we have

$$x = 36.$$

Therefore, there are 36 girls in Math Zoom Academy.

Problem 2.21 **Sally is playing basketball. After Sally takes 20 shots, she has made 55% of her shots. She takes 5 more shots and she raises her percentage to 56%. How many of the last 5 shots did she make?**

Answer

3

Solution 1

Sally made 55% of the first 20 shots, or

$$55\% \times 20 = 0.55 \times 20 = 11$$

shots. If she raises her percentage to 56% after 25 total shots, she must have made a total of

$$56\% \times 25 = 0.56 \times 25 = 14$$

shots. Hence she made

$$14 - 11 = 3$$

out of the last 5 shots.

Solution 2

(Algebra) Sally made 55% of the first 20 shots, or

$$55\% \times 20 = 0.55 \times 20 = 11$$

shots. Then let x be the number of shots she makes in the extra 5 shots. This means she makes

$$11 + x$$

of the 25 shots. We want this to be 56% of the shots, so

$$\frac{11 + x}{25} = 0.56.$$

Clearing denominators we get

$$11 + x = 14$$

so solving for x we have

$$x = 14 - 11 = 3.$$

Therefore, she made 3 out of the last 5 shots.

Problem 2.22 **Michelle has a dozen oranges and a dozen pears. Assume all the oranges are the same size and all the pears are the same size. Michelle uses her juicer to extract 8 ounces of pear juice from 3 pears and 8 ounces of orange juice from 2 oranges. She makes a pear-orange juice blend from an equal number of pears and oranges. What percent of the blend is pear juice?**

Answer

40%

Solution

We know that Michelle uses an equal number of pears and oranges. Since $2 \times 3 = 6$ it will be convenient if we assume she uses 6 pears and 6 oranges. Since 3 pears makes 8 ounces of juice and Michelle uses

$$2 \times 3 = 6$$

pears, she can make

$$2 \times 8 = 16$$

ounces of pear juice. Similarly, using

$$3 \times 2 = 6$$

oranges, she can make

$$3 \times 8 = 24$$

ounces of orange juice. The total amount of the juice is

$$16 + 24 = 40$$

ounces. Since 16 of those ounces are pear juice, the juice blend is

$$\frac{16}{40}0.4 = 40\%.$$

pear juice.

Problem 2.23 **The number of students going on a field trip changed from 24 to 36. What was the percentage increase in the number of students going on a field trip? Later, the number of students going changed back from 36 to 24. What was the percentage decrease in the number? Give both answers as a percentage rounded to the nearest whole number.**

Answer

Increase: 50%, Decrease: $\approx 33\%$

Solution

An increase from 24 to 36 students is a change of

$$36 - 24 = 12$$

students. Dividing by the original amount, this is a

$$12 \div 24 = 0.50 = 50\%$$

increase. Decreasing from 36 to 24 students is also a change of

$$36 - 24 = 12$$

students. However, the 'original' amount is now the 36 students, so the change is now a

$$12 \div 36 = 0.\overline{3} = 33.\overline{3} \approx 33\%$$

decrease.

Problem 2.24 **Sixty percent of the people on a subway train are seated. As some people prefer standing, only 75% of the seats on the subway are filled. If there are 12 empty seats, how many people are on the train?**

Answer

60

Solution 1

Since 75% of the seats are filled 25% of the seats are empty. Since there are a total of 12 empty seats, there must be

$$12 \div 25\% = 12 \div 0.25 = 48$$

seats in total. 75% of these seats are filled, so

$$75\% \times 48 = 0.75 \times 48 = 36$$

seats are filled. These 36 seats are filled by 36 people, and since 60% of the total people are seated, there must be a total of

$$36 \div 60\% = 36 \div 0.6 = 60$$

people on the train.

Solution 2

We know 75% of the seats are filled, so 25% of the seats are empty. Therefore the ratio of filled to empty seats is

$$75\% : 25\% = 75 : 25 = 3 : 1.$$

Since there are 12 empty seats, we multiply both sides of this ratio by 12 to get

$$3 : 1 = 36 : 12$$

so there are 36 filled seats. Since 60% of the people are seated on the train, 40% are standing. Therefore the ratio of seated to standing people is

$$60\% : 40\% = 60 : 40 = 3 : 2.$$

We know there are 36 seated people, so multiplying the ratio by

$$36 \div 3 = 12$$

we have

$$3 : 2 = 36 : 24$$

so there are 24 standing people. Thus, there are a total of

$$36 + 24 = 60$$

people on the train.

Problem 2.25 **Carson bought five notebooks from the College Bookstore at a cost of \$2.50 each. His brother Derick liked the notebooks and went to the bookstore the following day and also bought 5 notebooks. The bookstore had a 20%-off sale that day. How much did Derick save compared to Carson on the purchase of the five notebooks?**

Answer

$2.50

Solution 1

First we find the total amount that Carson paid,

$$5 \times 2.5 = 12.5$$

dollars. Derick also bought 5 notebooks, so without the sale he would also have paid \$12.50. Since Derick got 20% off, he had a discount of

$$20\% \times 12.50 = 0.20 \times 12.50 = 2.50$$

dollars. Therefore, Derick paid \$2.50 less than Carson.

Solution 2

Carson got 20% off, meaning he paid

$$100\% - 20\% = 80\%$$

of the full price. Since

$$80\% = 0.80 = \frac{4}{5}$$

this means that Carson paid the price of 4 notebooks to get 5 notebooks. Hence he saved the full price of

$$5 - 4 = 1$$

notebook, a total of $2.50.

Problem 2.26 A collector offers to buy the 1967 year of the sheep stamp sheet for 2000% of its face value. Bridget has one of the sheets with 12 stamps that had an original face value of 25 cents per stamp. How much would Bridget receive if she sold it to the collector?

Answer

$60

Solution

The collector is willing to pay 2000% of the original face value for each stamp. Therefore, they are willing to pay

$$2000\% \times 25 = 20 \times 25 = 500$$

cents or $5 for each stamp. Since Bridget's sheet contains 12 stamps, she can sell it for

$$12 \times 5 = 60$$

dollars to the collector.

Problem 2.27 The ratio of llamas to ostriches in the Math Zoom Academy petting zoo is $4 : 7$. If there are total of 44 llamas and ostriches in the petting zoo, how many of the them are llamas?

Answer

16

Solution 1

Since the ratio of llamas to ostriches is 4 : 7, 4 out of every

$$4 + 7 = 11$$

are llamas. Thus, $\dfrac{4}{11}$ of the 44 total of llamas and ostriches are llamas. Therefore, the petting zoo has

$$\frac{4}{11} \times 44 = 16$$

llamas.

Solution 2

(Algebra) Let the number of llamas be x, so the number of ostriches is $44 - x$. Since the ratio of llamas to ostriches is 4 : 7 we have the equation

$$\frac{4}{7} = \frac{x}{44 - x}.$$

Cross-multiplying we have

$$4 \times (44 - x) = 7x$$

so distributing and combining like terms we have

$$176 = 11x.$$

We then can solve for x to get

$$x = \frac{176}{11} = 16,$$

so there are 16 llamas at the petting zoo.

Problem 2.28 **Henry reads 160 pages of a book per day. After 5 days, Henry has $\dfrac{3}{5}$ of the book remaining. How many pages does the book have?**

Answer

2000

Solution

Henry reads 160 pages per day, so in 5 days he reads

$$5 \times 160 = 800$$

pages. Since Henry has $\dfrac{3}{5}$ of the book remaining the 800 pages he has read are $\dfrac{2}{5}$ of the book. Therefore, we can divide to get that the book has

$$800 \div \dfrac{2}{5} = 2000$$

pages in total.

Problem 2.29 **In a far-off land three fish can be traded for two loaves of bread and one loaf of bread can be traded for six ears of corn. How many ears of corn are worth the same as one fish?**

Answer

4

Solution

The ratio of fish to bread is $3 : 2$ and the ratio of bread to corn is $1 : 6$ in worth. Since
$$1 : 6 = 2 : 12$$
we can trade two loaves of bread for 12 ears of corn. Therefore, starting with twelve ears of corn, we can trade for two loaves of bread which we can then trade for three fish. Therefore, the ratio of corn to fish is $12 : 3$. Dividing by 3 we have
$$12 : 3 = 4 : 1$$
so four ears of corn are worth the same as one fish.

Problem 2.30 **Jenny starts with a full jar of jellybeans. Each day, she eats 20% of the jellybeans that were originally in the jar. At the end of the second day, 36 jellybeans remain. How many jellybeans were in the jar originally?**

Answer

60

Solution 1

Jenny eats 20% of the jellybeans each day, so after 2 days she has eaten

$$2 \times 20\% = 40\%$$

of the total jellybeans. This means 60% of the jellybeans remain, and we know this is equal to 36 jellybeans. To find the original amount, we divide

$$36 \div 60\% = 36 \div 0.6 = 60,$$

so there where originally 60 jellybeans in the jar.

Solution 2

(Algebra) Let x be the original amount of jellybeans. Jenny eats 20% of the jellybeans each day, so she eats

$$20\% \times x = 0.2x$$

Jellybeans per day. Since Jenny starts with x jellybeans and then eats $0.2x$ for two days to end up with 36 jellybeans remaining, we have the equation

$$x - 0.2x - 0.2x = 36.$$

Simplifying we have

$$0.6x = 36$$

so we get that

$$x = \frac{36}{0.6} = 60.$$

Therefore, there were originally 60 jellybeans in the jar.

Problem 2.31 **Two-thirds of the monkeys in a cage are seated in three-fourths of the spots. The rest of the monkeys are standing. If there are 6 empty spots, how many monkeys are in the cage?**

Answer

27

Solution

Three-fourths of the spots are taken, so one-fourth of them are empty. Since we know there are 6 empty spots, we divide to see that there must be

$$6 \div \frac{1}{4} = 24$$

total spots. Since three-fourths of the spots are occupied, there are

$$24 \times \frac{3}{4} = 18$$

monkeys seated. We know two-thirds of the total monkeys are seated, so dividing we get there are

$$18 \div \frac{2}{3} = 27$$

monkeys in the cage.

Problem 2.32 **A shopper buys a \$100 coat on sale for 20% off. An additional \$5 is taken off the sale price by using a discount coupon. A sales tax of 8% is paid on the final selling price. What is the total amount the shopper pays for the coat?**

Answer

\$81

Solution

Let's carry the calculation out step by step. First the shopper gets 20% off, so they must pay
$$100\% - 20\% = 80\%$$
of the total price, which is

$$100 \times 80\% = 100 \times 0.8 = 80$$

dollars. Taking off an additional $5 is

$$80 - 5 = \$75.$$

Tax is an additional 8%, so they must pay

$$100\% + 8\% = 108\%$$

of the $75. This gives a total amount of

$$75 \times 108\% = 75 \times 1.08 = 81$$

dollars the shopper pays for the coat.

Problem 2.33 **The table below gives the percent of students in each grade at school A and school B.**

	K	1	2	3	4	5	6
A	21%	12%	11%	15%	13%	17%	11%
B	18%	11%	16%	11%	13%	14%	17%

School A has 100 students, and school B has 200 students. If the two schools combined, what percent of the students are in grade 6?

Answer

15%

Solution

We know there are

$$100 + 200 = 300$$

total students. We need to know how many students are in grade 6. 11% of the 100 students in school A are in grade 6, so

$$11\% \times 100 = 11$$

students in school A are grade 6. Similarly,

$$17\% \times 200 = 34$$

students are in grade 6 from school B. This gives a total of

$$11 + 34 = 45$$

students in grade 6 if two schools combine. Therefore,

$$\frac{45}{300} = 0.15 = 15\%$$

of the students would be in grade 6 if the two schools combined.

Problem 2.34 Calvin bought four Avengers movie tickets for his friends at a cost of $12.50 each. His friend Mark wanted to watch the movie with his family as well and went to buy the same amount of tickets the following day. The theater had a 20%-off sale that day. How much did Mark save comparing to Karl on the purchase of four movie tickets?

Answer

$10

Solution

Since Calvin bought 4 tickets, he spend a total of

$$4 \times 12.50 = 50$$

dollars. With the 20%-off discount, Mark only needs to pay

$$100\% - 20\% = 80\%$$

of the price for each ticket. Therefore, Mark pays

$$80\% \times 12.50 = 0.80 \times 12.50 = 10$$

dollars per ticket. Since Mark also buys 4 tickets, he spends a total of

$$4 \times 10 = 40$$

dollars, a savings of

$$50 - 40 = 10$$

compared to the price Calvin pays.

Problem 2.35   In the popular TV show "Who Wants to be a Millionaire", contestants earn certain amount of money based on the number of questions they answer correctly. The dollar values of each question are shown in the following table (where k = 1000).

Question	1	2	3	4	5	6	7	8
Value	100	200	300	500	1k	2k	4k	8k
Question	9	10	11	12	13	14	15	
Value	16K	32K	64K	125K	250K	500K	1000K	

Between which two questions is the percentage increase of the value the smallest?

Answer

2 and 3

Solution

We start finding the percentage increase between questions. Between questions 1 and 2 there is a change of

$$200 - 100 = 100$$

dollars. Dividing by the original dollar amount (for question 1), we get that this is an increase of

$$100 \div 100 = 1 = 100\%.$$

Between questions 2 and 3 there is also a change of

$$300 - 200 = 100$$

dollars, which now is a percentage increase of

$$100 \div 200 = 0.50 = 50\%.$$

Using the same method between questions 3 and 4, we get a change of

$$500 - 300 = 200$$

divided by the original

$$200 \div 300 = 0.\overline{6} = 66.\overline{6}\%,$$

so there is roughly a 67% increase between questions 3 and 4. For the rest of the questions, the dollar value at least doubles (it doubles for every question except 11 to 12, where it more than doubles). Since all of these are at least a 100% increase, we get that the increase between questions 2 and 3 has the smallest percentage increase among all the questions.

Problem 2.36 **A middle school has 780 students, some of which go to Math Olympiad classes. Among those who attend Math Olympiad classes, $\dfrac{8}{17}$ are in 6th grade, and $\dfrac{9}{23}$ are in 7th grade. How many students do NOT attend Math Olympiad classes?**

Answer

389

Solution

The attendees of the Math Olympiad class in 6th and 7th grades combined are

$$\frac{8}{17} + \frac{9}{23} = \frac{337}{391}$$

of total attendees. Therefore the ratio of 6th and 7th graders to total number of students who attend Math Olympiad classes is $337 : 391$. Since this ratio is in lowest terms, we can multiply by 2 or 3 or ... to get

$$337 : 391 = 674 : 782 = 1011 : 1173 = \cdots.$$

Therefore, the total number of students who take Math Olympiad classes is either $391, 782, 1173, \ldots$. However, as there are only 780 students in the school, we know that 391 take Math Olympiad classes. Thus,

$$780 - 391 = 389$$

students do NOT attend Math Olympiad classes.

Problem 2.37 **Jim is paid an 8% commission on the first $800 of weekly sales, and a 14% commission on any sales past $800. If Jim's sales were $1300, what was his commission?**

Answer

$134

Solution

Jim earns 8% commission on the first $800 of sales, a total of

$$8\% \times 800 = 0.08 \times 800 = 64$$

dollars. He earns 14% on the additional

$$1300 - 800 = 500$$

dollars of sales, for an additional commission of

$$14\% \times 500 = 0.14 \times 500 = 70$$

dollars. In total, Jim's earns

$$64 + 70 = 134$$

dollars as commission.

Problem 2.38 **Linda receives a weekly salary of $120 plus a commission of 5% on all sales above $500 per week. During three weeks Linda's total sales were $1540, $1235, and $1040. What was her total paycheck for the three weeks?**

Answer

$475.75

Solution 1

For the first week, Linda receives commission for

$$1540 - 500 = 1040$$

dollars in sales, a total of

$$5\% \times 1040 = 0.05 \times 1040 = 52$$

dollars. Since she earns a base salary of $120 she earns

$$120 + 52 = 172$$

dollars the first week. For the second week she earns commissions on

$$1235 - 500 = 735$$

dollars in sales. Thus, in week two she earns

$$120 + 5\% \times 735 = 120 + 0.05 \times 735 = 120 + 36.75 = 156.75$$

dollars. Similarly, in the last week she earns

$$120 + 5\% \times (1040 - 500) = 120 + 0.05 \times 540 = 120 + 27 = 147$$

dollars. In total, Linda's paycheck is

$$172 + 156.75 + 147 = 475.75$$

dollars.

Solution 2

Linda receives a base salary of $120 dollars per week. For three weeks this is a total of

$$3 \times 120 = 360$$

dollars. Linda only receives commissions on her weekly sales about $500. In week one she earns commissions on

$$1540 - 500 = 1040$$

dollars in sales, in week two she earns commissions on

$$1235 - 500 = 735$$

dollars in sales, and in week three

$$1040 - 500 = 540$$

dollars in sales. Since she earns the same 5% commissions each week, we can calculate the total commission for

$$1040 + 735 + 540 = 2315$$

dollars in sales. This is

$$5\% \times 2315 = 0.05 \times 2315 = 115.75$$

dollars. Hence Linda's total paycheck is

$$360 + 115.75 = 475.75$$

dollars for the three weeks.

Problem 2.39 **The table shows some of the results of a survey. What percentage of the males surveyed read the newspaper?**

	Read	Don't Read	Total
Male	?	26	?
Female	58	?	96
Total	136	64	200

Answer

75%

Solution

Since 200 people were surveyed in total and we know 96 of them were female, a total of

$$78 + 26 = 104$$

males were surveyed. Similarly, a total of 136 people surveyed read the paper. Since 58 of them are female the other

$$136 - 58 = 78$$

were male. We can divide this by the total number of males

$$78 \div 104 = 0.75 = 75\%.$$

so see that 75% of the males surveyed read the newspaper.

Problem 2.40   **Rita has 36 marbles, 20 of which are red and 16 of which are white. Rose has 27 marbles, all of them either red or white. Suppose Rita and Rose have the same ratio of red to white marbles. How many more white marbles does Rita have than Rose?**

Answer

4

Solution 1

Rita has 20 red and 16 white marbles, so the ratio is

$$20 : 16 = 5 : 4$$

for both Rita and Rose. Therefore for every 4 white marbles Rose has she has 5 red marbles, so $\frac{4}{9}$ of Rose's marbles are white. We thus can calculate that Rose has

$$27 \times \frac{4}{9} = 12$$

white marbles. Since Rita has 16 white marbles, she has

$$16 - 12 = 4$$

more white marbles than Rose.

Solution 2

Rose has 27 marbles compared to the 36 marbles Rita has. Therefore Rose has

$$27 \div 36 = 0.75 = 75\%$$

of the marbles Rita has. Since they both have the same ratio of red to white marbles, this means Rosa will also have 75% of the white marbles Rita has, a total of

$$75\% \times 16 = 0.75 \times 16 = 12$$

white marbles. Hence Rita has

$$16 - 12 = 4$$

more white marbles than Rose.

Problem 2.41   **Three bags of jelly beans contain 26, 28, and 30 beans. The bags consist of respectively 50%, 25%, and 20% yellow jelly beans. All three bags of beans are dumped into one bowl. What percent of all beans are yellow jelly beans? Round your answer to the nearest percent.**

Answer

$\approx 31\%$

Solution

First find the number of yellow jellybeans in each bag. The first bag has

$$50\% \times 26 = 13,$$

the second has

$$25\% \times 28 = 7,$$

and the third has

$$20\% \times 30 = 6$$

yellow jelly beans. So the total number of yellow jelly beans in the three bags is

$$13 + 7 + 6 = 26.$$

The number of jelly beans of all colors is

$$26 + 28 + 30 = 84$$

so the percent of yellow jellybeans is

$$26 \div 84 = \frac{13}{42} \approx 0.3095 \approx 31\%.$$

Problem 2.42   **Andy had no money, so his Granny Smith gave him 36% of her money. Now Granny Smith has \$80 left, and Andy has \$2 more than Elberta. How many dollars does Elberta have?**

Answer

43

Solution

Granny Smith has $80 left after she gave 36% to Andy. This means the $80 is

$$100\% - 36\% = 64\%$$

of her starting money. Therefore before she gave Andy money Granny Smith had

$$80 \div 64\% = 80 \div 0.64 = 125$$

dollars. She gave Andy 36% of this, or

$$125 \times 36\% = 125 \times 0.36 = 45$$

dollars. Thus, Elberta has

$$45 - 2 = 43$$

dollars.

Problem 2.43 **Two 600 ml pitchers contain orange juice. One pitcher is 30% full and the other pitcher is 40% full. Water is added to fill each pitcher completely, then both pitchers are poured into one large container. What percent of the mixture in the large container is orange juice?**

Answer

35%

Solution 1

First we can find the amount of orange juice in each pitcher. The first pitcher contains

$$30\% \times 600 = 180$$

ml and the second contains

$$40\% \times 600 = 240$$

ml of orange juice. Then the total amount of orange juice is

$$180 + 240 = 420$$

ml. Since both pitchers are filled with water completely, then the total amount of liquid in the two pitchers is

$$600 + 600 = 1200$$

ml. Therefore the mixture in the large container is

$$420 \div 1200 = 0.35 = 35\%$$

orange juice.

Solution 2

If we fill both pitchers with water, the first is 30% orange juice and the second is 40% orange juice. Since both pitchers have the same size, the percentage of orange juice after mixing is the average of the two pitchers. Therefore the large contain contains

$$(30\% + 40\%) \div 2 = 35\%$$

orange juice.

Problem 2.44 **At a party there are only single women and married men with their wives. 40% of the women are single. What percentage of the people in the room are married men?**

Answer

30%

Solution

For convenience assume 100 women are at the party. Therefore

$$40\% \times 100 = 0.40 \times 100 = 40$$

of the women are single and

$$100 - 40 = 60$$

women are married. Thus there must also be 60 married men at the party. There are no single men at the party, so the total number of people is

$$40 + 60 + 60 = 160.$$

Hence the percentage of married men is

$$60 \div 160 = 3 \div 8 = 0.375 = 37.5\%$$

at the party.

Problem 2.45   **A and B are two identical cups. A is full with salt water containing 2% salt and B is half full with salt water containing 0.8% salt. Suppose we pour half of the salt water from cup A to cup B so cup B is now full of salt water. What percentage of salt is the salt water in cup B?**

Answer

1.4%

Solution 1

We do not know the size of the two cups, but all that matters is that they are the same size. For convenience let's assume each cup is 100 ml in volume.

For cup A, the amount of salt is

$$2\% \times 100 = 0.02 \times 100 = 2 \text{ ml}$$

and for cup B, the amount of salt is

$$0.8\% \times 50 = 0.008 \times 50 = 0.4 \text{ ml}.$$

Therefore half of cup A contains

$$2 \div 2 = 1 \text{ ml}$$

of the salt, so cup B ends up with

$$1 + 0.4 = 1.4 \text{ ml}$$

of salt. As cup B is now full, it contains 100 ml of salt water. Now, the percentage of salt in B is

$$1.4 \div 100 = \frac{7}{500} = 0.014 = 1.4\%.$$

Solution 2

When cup B is filled, it ends up with half of the salt water from cup A and half of the salt water from cup B. Cup A is 2% salt and cup B is 0.8% salt. Since both cups have the same size, there is an equal volume of salt water from both cups. Thus the percentage of salt after mixing is the average of the two cups. Therefore the salt water in cup B ends up being

$$(2\% + 0.8\%) \div 2 = 1.4\%$$

salt.

Problem 2.46 **Business is a little slow at Lou's Fine Shoes, so Lou decides to have a sale. On Friday, Lou increases all of Thursday's prices by 10%. Over the weekend, Lou advertises the sale: "Ten percent off the list price. Sale starts Monday." How much does one pair of shoes cost on Monday that cost \$40 on Thursday?**

Answer

$39.60

Solution

On Friday, the new price of the $40 shoes will be 10% higher than Thursday which is
$$(1 + 10\%) \times 40 = 1.10 \times 40 = 44$$
dollars. Then on Monday, the price is 10% of the $44 price, which is
$$(1 - 10\%) \times 44 = 0.90 \times 40 = 39.60.$$

Therefore, the shoes cost $39.60 on Monday.

Problem 2.47 **A merchant offers a large group of items at 30% off. Later, the merchant takes 20% off these sale prices and claims that the final price of these items is 50% off the original price. What is the true total discount?**

Answer

44%

Solution 1

The orignal price of the item is not given in the question. We can assign a price for it to help us with the calculation. Let's say the original price of the item is $100. With the first 30% off, the new price is

$$(1 - 30\%) \times 100 = 0.70 \times 100 = 70$$

dollars. With another 20% off, the price becomes

$$(1 - 20\%) \times 70 = 0.80 \times 70 = 56$$

dollars. This is a total discount of

$$100 - 55 = 44$$

dollars, so dividing by the original price the percentage discount is

$$44 \div 100 = 0.44 = 44\%.$$

So the merchant's claim is not correct and the true discount is 44% off the original price.

Solution 2

More generally, we can assume x is the original price. With the first 30% off, the sales price will be $0.7x$. With the next 20% off, the sales price will be $0.8 \times 0.7x = 0.56x$. Then compare the final price $0.56x$ with the original price x, we have

$$\frac{x - 0.56x}{x} = \frac{0.44x}{x} = 0.44.$$

So, the true total discount is 44%.

Problem 2.48 **A grocery store sells eggs in three sizes: small (S), medium (M) and large (L). The medium size costs 50% more than the small size and contains 20% fewer eggs than the large size. The large size contains twice as many eggs as the small size and costs 30% more than the medium size. Rank the three sizes from best to worst buy.**

Answer

Medium, Large, Small

Solution

The question asks to find out which size is the best to buy and which is the worst. This ranking is best obtained by calculating the unit price for each size.

We are not given any information about the actual prices or the actual amount of the eggs in each size. However, since all the relations between quantities are given in terms of percentage, we may assume a value for one of the quantities, and express everything else based on that value. We may treat the prices and the package size separately.

In terms of the price, let's assume the price of the small size is 100 cents. The medium size costs 50% more than the small one, so the price of medium size is

$$(1 + 50\%) \times 100 = 1.50 \times 100 = 150 \text{ cents.}$$

The large size costs 30% more than the medium one, so we can now calculate the price of the large size, which is

$$(1 + 30\%) \times 150 = 1.30 \times 150 = 195 \text{ cents.}$$

Now that we have the prices determined, let us work on the amounts of eggs each size contains. For simplicity, assume the small size contains 100 eggs. The large size contains twice as many eggs as the small one, so the large size contains

$$2 \times 100 = 200$$

eggs. The medium size contains 20% fewer eggs than the large size, so the medium size contains

$$(1 - 20\%) \times 200 = 0.80 \times 200 = 160$$

eggs.

We now have the prices and numbers of eggs for each size, we can calculate the unit prices per egg. The small size gives 100 eggs for 100 cents, so each egg costs

$$100 \div 100 = 1 \text{ cent.}$$

The medium size gives 160 eggs for 150 cents, so each egg costs

$$150 \div 160 = \frac{15}{16} = 0.9375 \text{ cent.}$$

Lastly, the large size gives 200 eggs for 195 cents, so each egg costs

$$195 \div 200 = \frac{39}{40} = 0.975 \text{ cent.}$$

For clarity we display the data in the following table.

Size	Price	Number	Unit Price
Small	100	100	1
Medium	150	160	0.9375
Large	195	200	0.975

From the table we see that the medium size has the best unit price and therefore the best buy. The small size has the worst unit price and is therefore the worst buy. The large size is the better than the small but worse than the medium to buy.

Problem 2.49 **A bag contains 3 blue, 4 red and 3 yellow marbles. How many blue marbles must be added to the bag for it to contain 75% blue marbles?**

Answer

18

Solution 1

(Trial and Error) Right now we have 3 blue marbles and total of

$$3 + 4 + 3 = 10.$$

marbles. Currently

$$3 \div 10 = \frac{3}{10} = 30\%$$

of the marbles are blue. We want the percentage of blue marbles to be

$$75\% = 0.75 = \frac{75}{100} = \frac{3}{4}.$$

Since the desired numerator in the ratio is 3 and we currently have 3 blue marbles, we try adding 3 blue marbles at a time. For example adding 3 more blue marbles we have a total of

$$3 + 3 = 6$$

blue marbles and

$$10 + 3 = 13$$

total marbles. Hence the new percentage of blue marbles is

$$6 \div 13 = \frac{6}{13} \approx 46\%.$$

We can continue in this manner adding more and more blue marbles until the percentage is 75%. The results are listed in the following table:

Blue Added	3	6	9	12	15	18
Blue Total	6	9	12	15	18	21
Total Marbles	13	16	19	22	25	28
Ratio	6/13	9/16	12/19	15/22	18/25	21/28
Percent	$\approx 46\%$	$\approx 56\%$	$\approx 63\%$	$\approx 68\%$	$\approx 72\%$	75%

Therefore, we should add 18 blue marbles to the bag.

Solution 2

Let us focus on the quantify that is **unchanged** in the process. Since blue marbles are to be added, the number of non-blue (red and yellow) marbles are unchanged. There are always 7 non-blue marbles to begin with, so the number of non-blue marbles remains at 7 after blue marbles are added. We want the blue marbles to be 75% of the total, thus non-blue marbles should be 25% of the total, which means there must be totally

$$7 \div 25\% = 7 \div \frac{1}{4} = 7 \times 4 = 28 \text{ marbles,}$$

and the number of blue marbles among these 28 is

$$28 \times 75\% = 28 \times \frac{3}{4} = 21.$$

Since we start with 3 blue marbles, we must add $21 - 3 = 18$ new blue marbles.

Solution 3

Let the number of blue marbles added to the bag to be x. After adding these marbles there are

$$3 + x$$

blue marbles and

$$3 + x + 4 + 3 = 10 + x$$

marbles in total. We want 75% of the marbles to be blue, so

$$\frac{3 + x}{10 + x} = 75\%,$$

which is the same as

$$\frac{3 + x}{10 + x} = \frac{3}{4}.$$

Cross multiplying we have

$$12 + 4x = 30 + 3x.$$

Combining like terms solves for x, with

$$x = 18.$$

Therefore, we need to add 18 blue marbles to the bag.

Problem 2.50 **Phil Lanthropist won a great deal of money in a contest. He gave 20% of his winnings to his parents, gave 25% of the remaining money to his children, and gave the remaining $900,000 to his favorite charity. What was the total number of dollars that Phil won?**

Answer

$\$1,500,000$

Solution 1

It is easiest to work backwards. He ends by giving $\$900,000$ to charity. This was Phil's remaining money after he gave 25% of the money he had left to charity. Hence the $\$900,000$ was

$$100\% - 25\% = 75\%$$

of the remaining money. Therefore he had

$$900,000 \div 75\% = 900,000 \div 0.75 = 1,200,000.$$

dollars before he gave his children money. Since he started by giving his parents 20% of his winnings, the $1,200,000$ was

$$100\% - 20\% = 80\%$$

of his winnings. Therefore we know his winnings was

$$1,200,000 \div 80\% = 1,500,000.$$

dollars.

Solution 2

Let the amount of Phil won to be x dollars. He gives 20% to his parents, so is left with

$$100\% - 20\% = 80\%$$

of his money. He then gives 25% of the remaining to his children, which is

$$100\% - 25\% = 75\%$$

of his remaining money. Therefore his remaining money is

$$75\% \times (80\% \times x)) = 0.75 \times 0.80 \times x = 0.6x.$$

He gives this remaining money to charity. Since he gave $900,000 to charity we know

$$0.6x = 900000$$

so solving for x we have

$$x = \frac{900000}{0.6} = 1500000.$$

Hence Phil won $1,500,000$ in the contest.

Problem 2.51 **In the fish tank at Albert's house, $\frac{1}{4}$ of the fish are red and the number of black fish is $\frac{3}{5}$ of the number of red fish. There are 24 additional fish that are all spotted. How many red fish are there?**

Answer

10

Solution 1

Since $\dfrac{1}{4}$ of the fish are red and the number of black fish is $\frac{3}{5}$ that of the number of red fish,

$$\frac{1}{4} \times \frac{3}{5} = \frac{3}{20}$$

of the total fish are black. Thus,

$$\frac{1}{4} + \frac{3}{20} = \frac{8}{20} = \frac{2}{5}$$

of the fish are either black or red. Hence, $\dfrac{3}{5}$ of the fish are spotted. There are 24 spotted fish, hence the total number of fish in the tank is

$$24 \div \frac{3}{5} = 40.$$

Recalling that $\dfrac{1}{4}$ of the fish are red we know there are

$$40 \times \frac{1}{4} = 10$$

red fish.

Solution 2

Let the total number of fish in the tank to be x. One-fourth are red, so there are

$$\frac{1}{4} \times x = \frac{x}{4}$$

red fish. There are $\dfrac{3}{5}$ths as many black fish, so there are

$$\frac{3}{5} \times \frac{x}{4} = \frac{3x}{20}$$

black fish. The only other type of fish is spotted. Thus, since there are 24 spotted fish we have

$$\frac{x}{4} + \frac{3x}{20} + 24 = x$$

Combining like terms we have

$$24 = \frac{3x}{5}$$

so we can solve for x to get

$$x = 24 \times \frac{5}{3} = 40,$$

the total number of fish. One-fourth are red, so the tank contains

$$\frac{1}{4} \times 40 = 10$$

red fish.

Problem 2.52 **The sales tax rate in Orange County is 8%. During a sale at an outlet in Orange County, the price of a suit is discounted 40% from its \$190.00 price. Two clerks, Jimmy and Tony, calculate the bill independently. Jimmy first adds 8% sales tax to the price, and then subtracts 40% from this total. Tony first subtracts 40% of the price from the original price, and then adds 8% sales tax to the discounted price. What is Jimmy's total minus Tony's total?**

Answer

0

Solution

Calculate each clerk's sale price, and then compare.

Jimmy first adds sales tax, which brings the total to

$$(1 + 8\%) \times 190 = 1.08 \times 190 = 205.20$$

dollars. He then gives the 40% discount. Thus,

$$(1 - 40\%) \times 205.2 = 0.60 \times 205.2 = 123.12.$$

dollars is the final price of the suit. On the other hand, Tony first applies the discount, giving a discounted price of

$$(1 - 40\%) \times 190 = 0.60 \times 190 = 114$$

dollars. He then adds the sales tax of 8%, for a final price of

$$(1 + 8\%) \times 114 = 1.08 \times 114 = 123.12$$

for the suit. Both totals are the same, so there is no difference between them.

Problem 2.53 **A fruit salad consists of crabapples, cranberries, black berries, and black cherries. If there are twice as many cranberries as crabapples and three times as many black berries as black cherries and four times as many black cherries as cranberries and the fruit salad has 280 total fruits, then how many black cherries does it have?**

Answer

64

Solution 1

To stay organized, let's list out all the ratios between each two fruits using the abbreviations CB = cranberries, CA = crabapples, BB = black berries, and BC = black cherries.

$$CB : CA = 2 : 1, BB : BC = 3 : 1, BC : CB = 4 : 1.$$

We can then find the ratios between all the fruits. We know the ratio of $CB : CA$ is $2 : 1$, and the ratio of $BC : CB$ is $4 : 1 = 8 : 2$, so the ratio $BC : CA$ is $8 : 1$ and the ratio of the three $CB : CA : BC = 2 : 1 : 8$. Similarly, since the ratio of $BB : BC$ is $3 : 1 = 24 : 8$ we have $CB : CA : BC : BB = 2 : 1 : 8 : 24$.

Since

$$2 + 1 + 8 + 24 = 35$$

the ratio above tells us that

$$\frac{8}{35}$$

of the fruits in the fruit salad are black cherries. Since the fruit salad has 280 total fruits, there are

$$\frac{5}{35} \times 280 = 64$$

black cherries in the fruit salad.

Solution 2

There are less crabapples than cranberries, less cranberries than black cherries, and less black cherries than black berries. Hence there are less crabapples than any other fruit. Suppose there was 1 crabapple in the fruit salad. Since there are twice as many cranberries, there would be

$$1 \times 2 = 2$$

cranberries. There are four times as many black cherries as cranberries, so there would be

$$2 \times 4 = 8$$

black cherries. Lastly, there are three times as many black berries as black cherries, so there would be

$$4 \times 3 = 24$$

black berries. Thus for every 1 crabapple there are 2 cranberries, 8 black cherries, and 24 black berries, a total of

$$1 + 2 + 8 + 24 = 35$$

fruit. Since

$$280 \div 35 = 8$$

if we multiple the amount of each fruit we have by 8 we will have the necessary 280 fruits in the fruit salad. In this case we will have

$$8 \times 8 = 64$$

black cherries.

Problem 2.54 **Jong-Zhi took a math test that had 12 arithmetic questions, 15 algebra questions and 18 geometry questions. She got 75% of the arithmetic questions correct and 60% of the algebra questions correct. How many of the geometry questions must she get correct to get a passing grade of 75%?**

Answer

18

Solution

The test in total has

$$12 + 15 + 18 = 45$$

questions. For Jong-Zhi to get 75% of them correct, she needs to get

$$75\% \times 45 = 0.75 \times 45 = 33.75$$

questions correct. Thus, to pass Jong-Zhi must get at least 34 questions correct. We know she got 75% of the arithmetic questions correct. Since there are 12 questions, she got

$$75\% \times 12 = 0.75 \times 12 = 9$$

arithmetic questions correct. Similarly, she got 60% of the 15 algebra questions correct which is an additional

$$60\% \times 15 = 9$$

correct questions. Therefore Jong-Zhi has answered

$$9 + 9 = 18$$

questions correct so far. Therefore she needs to answer

$$34 - 18 = 16$$

of the 18 geometry questions to get a passing grade on the test.

Problem 2.55 **A \$480 TV was put on sale for 30% off. It wasn't sold so the price was lowered an additional percent off the sale price making the new price \$285.60. What percentage was the second discount?**

Answer

15%

Solution

After the first 30% off, the price of the TV was

$$(1 - 30\%) \times 480 = 0.70 \times 480 = 336$$

dollars. Therefore the price decreased

$$336 - 285.60 = 50.40$$

dollars. Dividing by the $336 price we have this was a further discount of

$$50.10 \div 336 = 0.15 = 15\%.$$

Problem 2.56 **Chris's pay went from $20/hr to $25/hr after his first evaluation. After his second evaluation his pay was raised to $33/hr. What is the difference between the second raise as a percent and the first raise as as percent?**

Answer

7%

Solution

The first raise was a total of

$$25 - 20 = 5$$

dollars per hour. Dividing by the initial salary of $20 gives that this was a

$$5 \div 20 = 0.25 = 25\%$$

increase. The second raise was a total of

$$33 - 25 = 8$$

dollars per hour, an increase of

$$8 \div 25 = 0.32 = 32\%.$$

Therefore, the second raise was

$$32\% - 25\% = 7\%.$$

more than the first raise.

Problem 2.57 **John, Edward, and Dan did a fundraiser for the math club at school and raised a total of \$370. They divided the \$370 into three parts such that the second part is $\frac{1}{4}$ of the third part and the ratio between the first and the third part is $3 : 5$. Find the value of each part.**

Answer

First part \$120, second part \$50, third part \$200

Solution 1

In order to find each part, we first calculate the ratios between all three parts.

It is known that the second part is $\dfrac{1}{4}$ of the third part. This gives us the ratio of the second part to third part is $1 : 4$. We also know the ratio of the first to the third is $3 : 5$. To combine these ratios, multiply the first ratio by 5 to get the ratio of the second to third part is $1 : 4 = 5 : 20$. Similarly, multiplying the second ratio by 4 we have the ratio of the first to the third is $3 : 5 = 12 : 20$. Therefore, we have the ratio of the three parts, which is

$$\text{First} : \text{Second} : \text{Third} = 12 : 5 : 20.$$

The total of these three is

$$12 + 5 + 20 = 37.$$

Since

$$370 \div 37 = 10$$

we can multiply the ratio by 10 to get

$$\text{First} : \text{Second} : \text{Third} = 12 : 5 : 20 = 120 : 50 : 200$$

So the first part is \$120, the second part is \$50, and the last part is \$200. To double check, the total value is

$$120 + 50 + 200 = 370$$

dollars as needed.

Solution 2

Let x denote the value of the third part in dollars. The second part is one-fourth the value of the third part, so the second part has value

$$\frac{1}{4} \times x = \frac{x}{4}$$

dollars. The first part to the third part has ratio $3 : 5$, so the first part is $\frac{3}{5}$ths as valuable. Hence the first part has value

$$\frac{3}{5} \times x = \frac{3x}{5}$$

dollars. Since the three parts together are $370 we have

$$\frac{3x}{5} + \frac{x}{4} + x = 370.$$

Combining like terms we have

$$\frac{37x}{20} = 370.$$

Solve for x, then

$$x = \frac{20}{37} \times 370 = 200$$

dollars, the value of the third part. Then the second part is worth

$$\frac{1}{4} \times 200 = 50$$

dollars and the first part is

$$\frac{3}{5} \times 200 = 120$$

dollars.

Problem 2.58 Sally baked 5 dozen cookies on Saturday afternoon. She gave 60% of the cookies to her neighbors at the neighborhood barbecue. On Sunday, she took 75% of the remaining cookies to the church social. On Monday night, she and her family ate 50% of the remaining cookies while watching football. What percent of the 5 dozen cookies remain?

5%

Sally baked a total of

$$5 \times 12 = 60$$

cookies. After she gave 60% of the cookies, was left with

$$100\% - 60\% = 40\%$$

of the cookies, a total of

$$40\% \times 60 = 24$$

cookies. She then took 75% of these, leaving her with

$$100\% - 75\% = 25\%$$

or

$$25\% \times 24 = 6$$

cookies. Eating half of the cookies leave

$$50\% \times 6 = 3$$

remaining. Since Sally started with 60 cookies, she is left with

$$3 \div 60 = 0.05 = 5\%$$

of the cookies.

Problem 2.59 **In her history class, Marie averaged 90% correct on five 10 question quizzes, got 96% correct on a 50 question midterm exam, and 75% correct on an 80 question final exam. What is the percentage of correct answers she provided if the total points for correct answers on all quizzes and exams are combined?**

85%

Marie answered a total of

$$5 \times 10 + 50 + 80 = 50 + 50 + 80 = 180$$

questions in her history class. She answered 90% of the quiz questions correctly, a total of

$$90\% \times 50 = 45.$$

correct. For the midterm she answered

$$96\% \times 50 = 48$$

questions correct. Lastly, for the final she answered

$$75\% \times 80 = 60$$

questions correct. Therefore, the total number of correct questions Marie got correct in the class is

$$45 + 48 + 60 = 153,$$

and hence she got

$$153 \div 180 = \frac{17}{20} = 85\%$$

of the questions correct in total.

Problem 2.60 **The students in Miss Einstein's class took a math test. Two-thirds of the class passed and the other third failed. The ratio of boys to girls in the class is 2 to 1. All of the girls passed the exam. What percentage of boys failed the exam?**

Answer

50%

Solution 1

The question doesn't tell us how many girls or boys are in the class, so for convenience suppose there are 10 girls in the class. Since the ratio of boys to girls in the class is 2 : 1 there must be twice as many boys in the class, so there are 20 boys in the class. Thus, the class in total has

$$10 + 20 = 30$$

students. One-third of the class failed, so

$$\frac{1}{3} \times 30 = 10$$

students failed. All the girls passed, so these 10 students must all be boys. Since there are 20 boys in total,

$$10 \div 20 = \frac{1}{2} = 50\%$$

of the boys failed.

Solution 2

The ratio of boys to girls is 2 : 1, which tells us boys are $\frac{2}{3}$ of the class, and girls are the other $\frac{1}{3}$ of the class. We know that $\frac{2}{3}$ of the class passed and $\frac{1}{3}$ of the class failed. Since all the girls passed, the $\frac{1}{3}$ of the class that failed must all have been boys. Further, since $\frac{2}{3}$ of the class passed, and $\frac{1}{3}$ of the class was girls who passed the test,

$$\frac{2}{3} - \frac{1}{3} = \frac{1}{3}$$

of the class was boys who pased the test. Summarizing, $\frac{1}{3}$ of the class is boys who passed and $\frac{1}{3}$ of the class is boys who failed. This means there are the same number of boys who passed and boys who failed. Therefore, 50% of the boys failed the exam.

3. Chickens and Rabbits

Problem 3.1 There are some chickens and some rabbits on a farm. Suppose there are 45 heads and 128 feet in total among the chickens and rabbits, how many of the animals are chickens? How many are rabbits?

Answer

26 chickens, 19 rabbits

Solution 1

We use a creative method for this problem. Assume the chickens and rabbits are trained, and upon hearing a whistle, the chickens stand on one leg only and the rabbits stand on their two hind legs. Now you blow a whistle, the number of feet on the ground is reduced to half, which is

$$128 \div 2 = 64.$$

At this point, the chicken feet (on the ground) correspond one-to-one to their heads, and the rabbit feet are two-to-one to their heads. Each of the 45

---

heads contributes one foot, so every extra foot over 45 must come from a rabbit. This means there are

$$64 - 45 = 19$$

rabbits. The rest of the animals are chickens, so there are

$$45 - 19 = 26$$

chickens on the farm.

Solution 2

(Algebra) Assume there are x chickens and y rabbits. Every animal has (exactly) one head, so

$$x + y = 45.$$

Every chicken has 2 feet and every rabbit has 4 feet. Since there are 128 feet in total, we have

$$2 \times x + 2 \times y = 128.$$

Then we have the system of equations

$$\begin{cases} x + y &= 45, \\ 2x + 4y &= 128. \end{cases}$$

To solve for x and y, we can double the first equation to get

$$2x + 2y = 90$$

and subtract this equation from the second equation to get

$$2y = 38$$

and therefore

$$y = 19.$$

We can then substitute to solve for x to get

$$x = 45 - 19 = 26.$$

Hence, there are 26 chickens and 19 rabbits.

Problem 3.2 **In a math competition, there are 25 questions. Each correct answer earns 6 points. One point is taken away for each incorrectly answered or unanswered question. Jenny received 101 points. How many questions did she answer correctly?**

Answer

18

Solution 1

If Jenny answered all questions correctly, she would have gotten a perfect score of

$$25 \times 6 = 150.$$

However, her resulting score was 101. For each incorrect or blank answer, Jenny loses the 6 points that the she would have gotten for the question and an additional point for the incorrect answer. Thus, her score decreases by a total of

$$6 + 1 = 7$$

points for each question missed. The total decrease in the score was

$$150 - 101 = 49$$

points, and thus the number of incorrectly answered or unanswered question was

$$49 \div 7 = 7.$$

Therefore Jenny answered

$$25 - 7 = 18$$

questions correctly.

Solution 2

(Algebra) Let x be the number of questions Jenny answered correctly and y be the number Jenny answers incorrectly. There are 25 total questions, so

$$x + y = 25.$$

Jenny got 101 points total. Since every correct answer is worth 6 points and every incorrect answer is loses one point,

$$6 \times x - y = 101.$$

This gives us the system of equations

$$\begin{cases} x+y &=& 25, \\ 6x-y &=& 101. \end{cases}$$

Adding the two equations together we have

$$7x = 126$$

so we can solve for x to get

$$x = \frac{126}{7} = 18.$$

Therefore Jenny got 18 of the 25 questions correct.

Problem 3.3 **Two trucks dump dirt of 400 cubic meters. Truck A carries 7 cubic meters per load. Truck B carries 4 cubic meters per load. The dirt is removed after 70 loads. How many loads are carried by truck A?**

Answer

40

Solution 1

If we first assume all 70 loads are all from Truck B, there would be a total of

$$70 \times 4 = 280$$

cubic meters of dirt removed. In actuality,

$$400 - 280 = 120$$

more cubic meters of dirt were removed. This means that Truck A must have taken some of the loads of dirt. Each load from Truck A has

$$7 - 4 = 3$$

more cubic meters than that from Truck B, so if we take

$$120 \div 3 = 40$$

loads from truck B and give them to Truck A, we will have the correct amount of dirt. Hence truck A carries 40 loads of dirt.

Solution 2

(Algebra) Let x be the number of loads from Truck A and y be the number of loads from Truck B. There are 70 loads in total, so

$$x + y = 70.$$

Each load from Truck A takes 7 cubic meters of dirt, while every load from Truck B carries 4 cubic meters of dirt. Since there are 400 cubic meters of dirt in total,

$$7 \times x + 4 \times y = 400.$$

This gives us the system of equations

$$\begin{cases} x + y &= 70, \\ 7x + 4y &= 400. \end{cases}$$

Solving the first equation for y we have

$$y = 70 - x.$$

Substituting into the second equation we have

$$7x + 4 \times (70 - x) = 400.$$

Distributing and combining like terms we have

$$3x = 120$$

so solving for x we have

$$x = 40.$$

Hence the number of loads from Truck A is 40.

Problem 3.4 **The price of a pack of colored pencils is $19 and the price of a pack of regular pencils is $11. The math teacher bought 16 packs of pencils for a total of $280. How many packs of each type did he buy?**

Answer

13 packs of colored pencils, 3 packs of regular pencils

Solution 1

If all 16 packs were colored pencils, then the teacher would have spent

$$19 \times 16 = 304$$

dollars, which is

$$304 - 280 = 24$$

more dollars than the actual amount. Each pack of regular pencils is

$$19 = 11 - 8$$

dollars less expensive than colored pencils, and therefore the math teacher bought

$$24 \div 8 = 3$$

packs of regular pencils. The remaining

$$16 - 3 = 13$$

packs are colored pencils.

Solution 2

(Algebra) Suppose the math teacher buys x packs of colored pencils and y packs of regular pencils. He buys 16 packs in total so

$$x + y = 16.$$

The teacher spends \$280 in total. Since each pack of colored pencils is \$19 and each pack of regular pencils is \$11 we have

$$19 \times x + 11 \times y = 280.$$

Thus we have the system of equations

$$\begin{cases} x + y &= 16, \\ 19x + 11y &= 280. \end{cases}$$

Multiplying the first equation by 19 we have

$$19x + 19y = 304$$

and subtracting the second equation from this gives

$$8y = 24.$$

Thus,

$$x = \frac{24}{8} = 3.$$

Substituting into the first equation we have

$$x + 3 = 16$$

so

$$x = 16 - 3 = 13.$$

Hence the math teacher buys 13 packs of colored pencils and 3 packs of regular pencils.

Problem 3.5 **Sarah counts her chickens and rabbits, and there are 16 heads and 44 feet. How many of each type are there?**

Answer

10 chickens, 6 rabbits

Solution 1

There are 16 animals in total. Imagine first all the animals are chickens. Then there would be

$$16 \times 2 = 32$$

total feet, which is

$$44 - 32 = 12$$

less than the actual number of feet. Since each rabbit has 2 more feet than a chicken, we need to replace

$$12 \div 2 = 6$$

chickens with rabbits. Therefore, Sarah has 6 rabbits and

$$16 - 6 = 10$$

chickens.

Solution 2

(Algebra) Let x be the number of chickens Sarah has and y be the number of rabbits. Since there are 16 heads in total we have

$$x + y = 16.$$

Each chicken has 2 feet and each rabbit has 4 feet, so since we know there are 44 feet in total we have

$$2 \times x + 4 \times y = 44.$$

This gives the system of equations

$$\begin{cases} x + y &=& 16, \\ 2x + 4y &=& 44. \end{cases}$$

Multiplying the first equation by 2 we have

$$2x + 2y = 32$$

and subtracting this from the second equation we get

$$2y = 12.$$

Hence we know

$$y = \frac{12}{2} = 6.$$

Substituting into the first equation,

$$x + 6 = 16$$

so

$$x = 16 - 6 = 10.$$

Thus we have found that Sarah has 10 chickens and 6 rabbits.

Problem 3.6 **In a farm there are goats and ducks. The total number of heads is 100, and the total number of legs is 316. How many animals of each type are there?**

Answer

58 goats, 42 ducks

Solution 1

Using our imagination, pretend all the ducks stand on one foot and all the goats stand up on their back 2 legs. There are still 100 heads so 100 animals, but now only half of the legs,

$$316 \div 2 = 158,$$

of the legs are on the ground. Each animal has one leg on the ground, and each goat has one extra leg on the ground. Since each duck now has one leg on the ground, the extra

$$158 - 100 = 58$$

legs come from goats. Thus there are 58 goats and

$$100 - 58 = 42$$

ducks on the farm.

Solution 2

(Algebra) Let x be the number of goats on the farm and y be the number of ducks. Since each animal has one head,

$$x + y = 100.$$

Each goat has 4 legs while each duck has 2 legs. There are a total of 316 legs on the farm, so

$$4 \times x + 2 \times y = 316.$$

Hence we have the system of equations

$$\begin{cases} x + y & = & 100, \\ 4x + 2y & = & 316. \end{cases}$$

Dividing the second equation by 2 we get

$$2x + y = 158.$$

Subtracting the first equation from this we have

$$x = 58,$$

the number of goats. Substituting into the first equation,

$$58 + y = 100,$$

so we have

$$y = 42,$$

the number of ducks on the farm.

Problem 3.7 **Sixty vehicles (cars and motorcycles) are parked in a parking lot. Totally there are 190 wheels. Given that a car has 4 wheels and a motorcycle has 2 wheels, how many cars and motorcycles each are in the parking lot?**

Answer

35 cars, 25 motorcycles

Solution 1

If all sixty vehicles where motorcycles, we would have a total of

$$60 \times 2 = 120$$

whells in the parking lot. We know there are actually 190 wheels in the lot,

$$190 - 120 = 70$$

more wheels than if we only have motorcycles. Switching a motorcycle for a car gives 2 extra wheels, therefore we must replace

$$70 \div 2 = 35$$

motorcycles with cars to get the correct number of wheels in the parking lot. Hence there are 35 cars and

$$60 - 35 = 25$$

motorcycles in the parking lot.

Solution 2

(Algebra) Let x be the number of cars in the parking lot and y the number of motorcycles. There are 60 vehicles in total, so

$$x + y = 60.$$

Each car has 4 wheels and each motorcycle has 2, so

$$4 \times x + 2 \times y = 190$$

is the total number of wheels in the lot. This gives the system of equations

$$\begin{cases} x + y & = & 60, \\ 4x + 2y & = & 190. \end{cases}$$

Doubling the first equation we have

$$2x + 2y = 120.$$

Subtracting this from the second equation we have

$$2x = 70,$$

so we solve for x to get

$$x = \frac{70}{2} = 35.$$

Substituting into the first equation,

$$35 + y = 60$$

so

$$y = 60 - 35 = 25.$$

There are thus 35 cars and 25 motorcycles in the parking lot.

Problem 3.8 **Teachers and students from the Areteem Summer Camp visited the museum. They bought a total of 99 tickets for 218 dollars. If each teacher ticket costs 4 dollars, and each student ticket costs 2 dollars, how many teachers and students were there respectively?**

Answer

10 teachers, 89 students

Solution 1

If instead we have 99 unsupervised students with no teachers, the 99 tickets would cost a total of

$$99 \times 2 = 198$$

dollars. This is

$$218 - 198 = 20$$

dollars cheaper than the actual 99 tickets. If we remove one student ticket and add one teacher ticket, it costs $2 extra. Since $20 extra was paid over the price of 99 student tickets, there must be

$$20 \div 2 = 10$$

teachers. Hence there are

$$99 - 10 = 89$$

students.

Solution 2

(Algebra) Let x be the number of teachers and y be the number of students who visit the Amusement Park. 99 tickets are bought, so

$$x + y = 99.$$

A total of $218 is spent buying these tickets. Each teacher ticket costs $4 and each student tickets costs $2, so

$$4 \times x + 2 \times y = 218.$$

We then need to solve the system of equations:

$$\begin{cases} x + y & = & 99, \\ 4x + 2y & = & 218. \end{cases}$$

Doubling the first equation we have

$$2x + 2y = 198.$$

Subtracting this from the second equation we have

$$2x = 20,$$

so we solve for x to get

$$x = \frac{20}{2} = 10.$$

Substituting into the first equation,

$$10 + y = 99$$

so

$$y = 99 - 10 = 89.$$

There are thus 10 teachers and 89 students who go to the Amusement Park.

Problem 3.9 The counselor brought his 51 students to the lake to go rowing, 6 people for each big boat and 4 people for each small boat. They rented 11 boats to fit everyone with no empty seats. How many big and small boats each did they rent?

Answer

4 big boats, 7 small boats

Solution 1

Remember that the counselor is a person too! This means there are a total of 52 people who fit in the boats. With 11 small boats, there is only room for

$$11 \times 4 = 44$$

people, so the counselor must have rented some big boats. The counselor needs room for

$$52 - 44 = 8$$

more people. Since a big boat carries

$$6 - 4 = 2$$

more people than a small boat, switching

$$8 \div 2 = 4$$

of the small boats to big boats will make sure the 11 boats fit everyone. This means that the counselor rented 4 big boats and

$$11 - 4 = 7$$

small boats.

(Algebra) Let x be the number of big boats and y be the number of small boats that the counselor rented. We know there are 11 boats in total, so

$$x + y = 11.$$

We need room for 51 students and 1 counselor, so 52 people in total. Each big boat carries 6 and each small boat carries 4, so

$$6 \times x + 4 \times y = 52.$$

We then need to solve the system of equations

$$\begin{cases} x + y & = & 11, \\ 6x + 4y & = & 52. \end{cases}$$

Multiplying the first equation by 4 we have

$$4x + 4y = 44.$$

Subtracting this equation from the second equation gives us

$$2x = 52 - 44 = 8$$

so solving for x we have

$$x = 4.$$

We can then substitute into the first equation to get

$$4 + y = 11$$

so solving for y we have

$$y = 11 - 4 = 7.$$

Hence we know that 4 big boats and 7 small boats were rented.

Problem 3.10 **Each set of chess is played by 2 students, and each set of Chinese checkers is played by 6 students. A total of 26 sets of chess and Chinese checkers are played by exactly 120 students in a school event. How many sets of each game are there?**

Answer

9 sets of chess, 17 sets of Chinese checkers

Solution 1

Two students can play with one set of chess, so if all 26 sets are chess sets, then only

$$26 \times 2 = 52$$

students can participate in the school event. Every set we switch from chess to Chinese checkers allows for

$$6 - 2 = 4$$

more students to participate. We know 120 students played in the event, which is

$$120 - 52 = 68$$

more than can play with only chess sets. Therefore we need to swap

$$68 \div 4 = 17$$

of the chess sets for Chinese checkers sets. Hence there are

$$26 - 17 = 9$$

chess sets and 17 Chinese checkers sets at the event.

Solution 2

(Algebra) Let x be the number of chess sets and y be the number of Chinese checkers sets at the event. Since there are 26 sets in total,

$$x + y = 26.$$

A chess set is played by 2 students and a Chinese checkers set is played by 6, so since 120 students in total play one of the two games we have that

$$2 \times x + 6 \times y = 120.$$

We are left to solve the system of equations

$$\begin{cases} x+y & = & 26, \\ 2x+6y & = & 120. \end{cases}$$

Doubling the first equation gives us

$$2x+2y = 52.$$

Subtracting from the second equation, this gives us

$$4y = 68$$

so solving for y,

$$y = \frac{68}{4} = 17.$$

We then use the first equation,

$$x+17 = 26,$$

so solve for x,

$$x = 26 - 17 = 9.$$

Hence there are 9 chess sets and 17 Chinese checkers sets used at the event.

Problem 3.11 Use 400 **matches to make triangles and pentagons. Totally** 88 **triangles and pentagons are made with no matches left over. How many of each shape are made?**

Answer

68 pentagons, 20 triangles

Solution 1

Every triangle has 3 sides, so needs 3 matches. To make 88 triangles we then need

$$88 \times 3 = 264$$

matches. A pentagon has 5 sides, so using

$$5 - 3 = 2$$

extra matches we can turn one of the triangles into a pentagon. There are

$$400 - 264 = 136$$

extra matches, so we can turn

$$136 \div 2 = 68$$

triangles into pentagons. Thus there are

$$88 - 68 = 20$$

triangles and 68 pentagons.

Solution 2

(Algebra) Let x be the number of triangles and y be the number of pentagons. There are 88 shapes in total so

$$x + y = 88.$$

A triangle has 3 sides so needs 3 matches. Similarly a pentagon has 5 sides so needs 5 matches. Since 400 matches are used in total,

$$3 \times x + 5 \times y = 400.$$

Hence we are left to solve the equation

$$\begin{cases} x + y & = & 88, \\ 3x + 5y & = & 400. \end{cases}$$

Multiplying the first equation by 3 gives us

$$3x + 3y = 264.$$

Subtracting this from the second equation,

$$2y = 136,$$

so

$$y = \frac{136}{2} = 68.$$

Substituting into the first equation,

$$x + 68 = 88$$

so

$$x = 88 - 68 = 20.$$

Thus there are 20 triangles and 68 pentagons.

Problem 3.12 There are 20 questions in a math competition. Five points are given to each correct answer, and -3 points are for each incorrect answer or unanswered question. Jeff received 60 points in the competition. How many questions did he answer correctly?

Answer

15

Solution 1

Since each question is worth 5 points, if Jeff got every question correct he would have received a score of

$$20 \times 5 = 100.$$

For every question Jeff gets wrong, he loses the 5 points for a correct answer and an additional 3 points for the incorrect answer, a total decrease of

$$5 + 3 = 8$$

points. Since his actual score is

$$100 - 60 = 40$$

points less than a perfect score, this means Jeff answered

$$40 \div 8 = 5$$

questions incorrectly. Hence he got

$$20 - 5 = 15$$

questions correct in the competition.

Solution 2

(Algebra) Let x be the number of questions Jeff got correct in the competition. There are 20 total questions, so this means Jeff got

$$20 - x$$

questions incorrect. He gets 5 points per correct answer and loses 3 points per incorrect answer. Since his total score was 60, we have

$$5 \times x - 3 \times (20 - x) = 60.$$

Distributing and combining like terms we have

$$8x = 120.$$

Hence

$$x = \frac{120}{8} = 15$$

so Jeff got 15 questions correct.

Problem 3.13 **Tiffany scored 29 points in her school's playoff basketball game. She made a combination of 2-point shots and 3-point shots during the game. If she made a total of 11 shots, how many 3-point shots did she make?**

Answer

7

Solution 1

If all of Tiffany's shots were 2-point shots, she would have scored

$$11 \times 2 = 22$$

points. She actually scored

$$29 - 22 = 7$$

more points. Each 3-point shot is worth an extra point over a 2-point shot, so 7 of Tiffany's shots must have been 3-point shots.

Solution 2

(Algebra) Let x be the number of 2-point shots and y be the number of 3-point shots Tiffany makes. She takes a total of 11 shots, so

$$x + y = 11.$$

These two types shots are worth 2 points and 3 points, so since Tiffany scores 29 points in total,

$$2 \times x + 3 \times y = 29.$$

This gives us the system of equations

$$\begin{cases} x + y & = & 11, \\ 2x + 3y & = & 29. \end{cases}$$

Solving the first equation for x gives us

$$x = 11 - y.$$

Substituting this into the second equation we have

$$2 \times (11 - y) + 3y = 29,$$

which after distributing and combining like terms gives us

$$y = 29 - 22 = 7.$$

So Tiffany made 7 baskets of the 3-point baskets.

Problem 3.14 **There are 48 tables in a restaurant. Small tables can seat 2 people, and big tables can seat 5 people. They can accommodate a maximum number of 159 people. How many small tables and how many big tables are there?**

Answer

27 small, 21 big tables

Solution 1

If all 48 tables are small tables, the restaurant could fit a maximum of

$$48 \times 2 = 96$$

people. This is

$$159 - 96 = 63$$

fewer people than the restaurants actual capacity. Each big table can accommodate

$$5 - 2 = 3$$

more people than the small table. Now

$$63 \div 3 = 21,$$

so if we switch 21 of the small tables to big tables the restaurant can accommodate 159 people as needed. Hence there are

$$48 - 21 = 27$$

small tables and 21 big tables at the restaurant.

Solution 2

(Algebra) Let x be the number of small tables and y be the number of big tables at the restaurant. There are 48 tables in total, so

$$x + y = 48.$$

The restaurant can accommodate 159 people, so since 2 people fit at a small table and 5 people fit at a big table,

$$2 \times x + 5 \times y = 159.$$

We then need to solve the system of equations

$$\begin{cases} x + y & = & 48, \\ 2x + 5y & = & 159. \end{cases}$$

Doubling the first equation,

$$2x + 2y = 96.$$

Subtracting this from the second equation we get

$$3y = 63.$$

Solving for y,

$$y = 21,$$

the number of big tables. Then we can substitute this in the first equation,

$$x + 21 = 48,$$

so the number of small tables is

$$x = 48 - 21 = 27.$$

Problem 3.15 **In Bob's Cycle Shop, workers received the delivery of an order Bob placed for the single seat bicycles and tricycles. There are a total of 90 seats and 215 wheels, plus all the other necessary parts and accessories. How many bicycles and how many tricycles can they assemble?**

Answer

55 bicycles, 35 tricycles

Solution 1

If the workers at the shop build 90 bicycles from the parts, they will end up using all 90 seats but only

$$90 \times 2 = 180$$

wheels, so they would have

$$215 - 180 = 35$$

wheels left over. Each tricycle has 1 more wheel than a bicycle, so the workers need to turn 35 of the bicycles into tricycles. Thus they assemble

$$90 - 35 = 55$$

bicycles and 35 tricycles.

Solution 2

(Algebra) Let x be the number of bicycles and y be the number of tricycles. Each has one seat, so
$$x + y = 90.$$
Each bicycles has 2 and each tricycles has 3 wheels, so

$$2 \times x + 3 \times y = 215.$$

This gives us the system of equations

$$\begin{cases} x + y & = & 90, \\ 2x + 3y & = & 215. \end{cases}$$

Doubling the first equation we have

$$2x + 2y = 180,$$

and subtracting this from the second we have

$$y = 215 - 180 = 35.$$

Substituting this into the first equation gives us

$$x + 35 = 90$$

so we can solve for x to get

$$x = 90 - 35 = 55.$$

Thus there are 55 bicycles and 35 tricycles.

Problem 3.16   **The company Green Pilots is organizing a company picnic at the beach. They want to save energy by carpooling to the beach site. They are able to fit 450 people into 80 vehicles that are either 5-seat sedans or 7-seat minivans. How many sedans and how many minivans do they need to use to take everyone to the picnic?**

Answer

55 sedans, 25 minivans

Solution 1

If the company uses 80 minivans, they can transport

$$80 \times 7 = 560$$

people to the beach. This is

$$560 - 450 = 110$$

more than they need to take, so they can switch some of the minivans for more energy efficient sedans. Each sedan holds

$$7 - 5 = 2$$

less people than a minivan. Therefore, they can switch

$$110 \div 2 = 55$$

minivans for sedans and still take everyone to the picnic. In the end Green Pilots can use 55 sedans and

$$80 - 55 = 25$$

to take everyone to the beach.

Solution 2

(Algebra) Let x be the number of 5-seat sedans and y be the number of 7-seat minivans Green Pilots uses. There are 80 total vehicles, so

$$x + y = 80.$$

The sedans fit 5 people and the minivans fit 7. There are 450 people in total, so we also know that

$$5x + 7y = 450.$$

This gives us the system of equations

$$\begin{cases} x + y &= 80, \\ 5x + 7y &= 450. \end{cases}$$

Multiplying the first equation by 7, we have

$$7x + 7y = 560$$

and subtracting the second equation from this gives

$$2x = 110.$$

Then we solve for x,

$$x = \frac{110}{2} = 55.$$

Plugging this into the first equation,

$$55 + y = 80,$$

and thus

$$y = 80 - 55 = 25.$$

Hence Green Pilots uses 55 sedans and 25 minivans to take everyone to the beach.

Problem 3.17 **The Math Zoom Camp allows students to form four-person teams or six-person teams to work together on projects. If there are 42 teams formed with the 200 total students in the camp, how many four-person teams and six-person teams are formed?**

Answer

26 four-person, 16 six-person teams

Solution 1

Start by creating 42 teams with 4 people per team. This uses

$$42 \times 4 = 168$$

of the 200 students, leaving

$$200 - 168 = 32$$

students left over. Since

$$6 - 4 = 2,$$

if we divide the 32 leftover students into

$$32 \div 2 = 16$$

pairs, we can combine these pairs with the four-person teams to get 16 six-person teams and

$$42 - 16 = 26$$

four-person teams.

Solution 2

(Algebra) Let x be the number of four-person teams and y be the number of six-person team. There are 42 teams in total, so

$$x + y = 42.$$

We also have the 200 students are divided into teams of four and six, so

$$4 \times x + 6 \times y = 200.$$

Multiplying the first equation by 4 gives us

$$4x + 4y = 168.$$

Subtracting this from the second equation we have

$$2y = 32,$$

and thus that

$$y = \frac{32}{2} = 16.$$

Going back to the first equation, we then have

$$x + 16 = 42,$$

so

$$x = 42 - 16 = 26.$$

Therefore the 200 students are divided into 26 four-person teams and 16 six-person teams.

Problem 3.18   **17 people went to a farm. There were goats to feed and chickens to feed, and each person fed exactly one animal. It costs \$1.50 to feed a chicken and \$2.00 to feed a goat. In total, the people spent \$32.50. How many of each type of animal did they feed?**

Answer

3 chickens, 14 goats

Solution 1

If all 17 people fed goats, they would spend a total of

$$17 \times 2 = 34$$

dollars. To save some money, some people fed chickens instead. Each chicken costs

$$2 - 1.5 = 0.5$$

dollars less than a goat. Since the people spend

$$34 - 32.5 = 1.5$$

dollars less than \$34,

$$1.5 \div 0.5 = 3$$

of the people fed goats. This means

$$17 - 3 = 14$$

fed chickens.

Solution 2

(Algebra) Let x be the number of chicken, and y be the number of goats. There are 17 animals fed in total, so

$$x + y = 17.$$

Since the people spend \$32.50 in total, we know

$$1.5 \times x + 2 \times y = 32.50.$$

Simplifying we have the system of equations

$$\begin{cases} x + y & = & 17, \\ 1.5x + 2y & = & 32.50. \end{cases}$$

Doubling the first equation,

$$2x + 2y = 34.$$

Subtracting the second equation from this we have

$$0.5x = 1.5$$

so multiplying by 2 on both sides we have

$$x = 2 \times 1.5 = 3.$$

Substituting into the first equation,

$$3 + y = 17,$$

so

$$y = 17 - 3 = 14.$$

Hence the people fed 3 chickens and 14 goats.

Problem 3.19 **A large family of 20 people goes to a restaurant. They each order either pizza or salad. The pizza costs $9.00 and salad costs $3.00. In all the family spends $138.00. How many pizzas and how many salads did the family order?**

Answer

7 salads, 13 pizzas

Solution 1

Had the whole family ordered salad, they would have spent a total of 60 dollars, which was

$$138 - 60 = 78$$

dollars less than they actually spent. Switching from salad to pizza cost an extra

$$9 - 3 = 6$$

dollars. Therefore, if

$$78 \div 6 = 13$$

people switched their order from salad to pizza the family would spend a total of $138. Hence

$$20 - 13 = 7$$

ordered salad and 13 ordered pizza.

Solution 2

(Algebra) Let x be the number of salads and y be the number of pizzas. Each person in the family orders one food, so

$$x + y = 20.$$

Salad costs $3 and pizza $9, so

$$3 \times x + 9 \times y = 138$$

because the family spend $138 in total. This gives the system of equations

$$\begin{cases} x + y & = & 20, \\ 3x + 9y & = & 138. \end{cases}$$

First we can simplify the second equation by dividing both sides by 3,

$$x + 3y = 46.$$

Subtracting the first equation from the new one gives us

$$2y = 26$$

so

$$y = \frac{26}{2} = 13.$$

Using this in the first equation we have

$$x + 13 = 20$$

so

$$x = 20 - 13 = 7$$

and therefore the family orders 7 salads and 13 pizzas.

Problem 3.20 **A pet owner has cats and birds. There are 25 pets in total and all combined the pets have a total of 90 legs. How many of each are there?**

Answer

5 birds, 20 cats

Solution 1

If all the pets were cats, there would be a total of

$$25 \times 4 = 100$$

legs, which is

$$100 - 90 = 10$$

more than the actual number. Each bird has

$$4 - 2 = 2$$

fewer legs than a cat. Therefore, as

$$10 \div 2 = 5$$

if we replace 5 cats with birds we reduce the number of legs by 10 as needed. Hence the owner has 5 birds and

$$25 - 5 = 20$$

cats.

Solution 2

(Algebra) Let x be the number of birds and y be the number of cats. There are 25 pets in total, so

$$x + y = 25.$$

There are 90 legs in total, so counting them separately for birds and cats we have

$$2 \times x + 4 \times y = 90.$$

We then need to solve the system of equations

$$\begin{cases} x + y & = & 25, \\ 2x + 4y & = & 90. \end{cases}$$

Multiplying the first equation by 4 we have

$$4x + 4y = 100.$$

If we then subtract the second equation from this we get that

$$2x = 10$$

so

$$x = \frac{10}{2} = 5.$$

Substituting back into the first equation,

$$5 + y = 25$$

so

$$y = 25 - 5 = 20.$$

Hence the pet owner has 5 birds and 20 cats.

Problem 3.21 **Sasha takes a mathematics competition. There are a total of 20 problems. For each correct answer, competitors receive 5 points. For each wrong answer, they instead get 1 point taken away. Sasha has 64 points in total. How many problems did she answer correctly?**

Answer

14

Solution 1

If Sasha answered all 20 questions correct she would receive a perfect score of

$$20 \times 5 = 100.$$

She actually has only 64 points, so

$$100 - 64 = 36$$

points were deducted. Since 1 point is taken away for each wrong answer, each wrong answer actually decreases the score by

$$1 + 5 = 6$$

points. The number of wrong answer is then

$$36 \div 6 = 6,$$

and hence

$$20 - 6 = 14$$

is the number of questions Sasha answered correctly.

Solution 2

(Algebra) Let x be the number of correct answers and y be the number of wrong answers. There are 20 questions in total, each correct or wrong so

$$x + y = 20.$$

The final score of 64 is a combination of 5 points for each correct answer and losing one point for each incorrect answer, so

$$5 \times x - 1 \times y = 64.$$

This gives the system of equations:

$$\begin{cases} x + y &= 20, \\ 5x - y &= 64. \end{cases}$$

Adding these two equations we have that

$$6x = 84$$

so dividing by 6 gives us

$$x = \frac{84}{6} = 14,$$

the number of questions Sasha got correct.

Problem 3.22 **Aria has 16 coins that are nickels and dimes. The total value is $1.20. How many nickels and dimes does Aria have?**

Answer

8 nickels, 8 dimes

Solution 1

It will be easiest if we convert everything from dollars to cents. Aria has a total of 1.2 dollars or

$$1.2 \times 100 = 120$$

cents. Assuming all 16 coins are nickels, there would be total of

$$16 \times 5 = 80$$

cents,

$$120 - 80 = 40$$

cents less than the actual amount Aria has. Each dime is worth

$$10 - 5 = 5$$

more cents than a nickel, so trading one nickel for a dime results in an increase of 5 cents. Therefore replacing

$$40 \div 5 = 8$$

nickels with dimes, we get the correct value of 120 cents. Hence Aria has 8 dimes and the other

$$16 - 8 = 8$$

coins are nickels.

Solution 2

(Algebra) Let x be the number of nickels and y be the number of dimes. For convenience we work with cents instead of dollars. Aria has 16 coins in total, so

$$x + y = 16.$$

Since Aria has nickels and dimes worth 1.2 dollars or 120 cents,

$$5 \times x + 10 \times y = 120.$$

If we solve the system of equations

$$\begin{cases} x + y & = & 16, \\ 5x + 10y & = & 120. \end{cases}$$

we will then know how many of each coin Aria has. Multiplying the first equation by 5 gives us

$$5x + 5y = 80$$

and subtracting from the second we have

$$5y = 120 - 80 = 40.$$

Solving for y,

$$y = \frac{40}{5} = 8.$$

Substituting back into the first equation,

$$x + 8 = 16,$$

so

$$x = 16 - 8.$$

Aria hence has 8 nickels and 8 dimes.

Problem 3.23   **There are 100 birds and cats. The birds have 80 more legs than the cats. How many birds and how many cats are there?**

Answer

80 birds, 20 cats

There are 80 more bird legs than cat legs. Since each bird has 2 legs, if we remove

$$80 \div 2 = 40$$

birds then we are left with

$$100 - 40 = 60$$

animals and there are an equal number of bird and cat legs. Since each cat has twice as many legs as a bird, there must be half as many birds as cats. Therefore there are

$$60 \div 3 = 20$$

cats and

$$20 \times 2 = 40$$

of the remaining animals are birds. Hence there are

$$40 + 40 = 80$$

birds in total.

(Algebra) Let x be the number of birds and y be the number of cats. There are 100 animals in total, so

$$x + y = 100.$$

Each bird has 2 legs and each cat 4 legs, so since there are 80 more bird legs than cat legs,

$$2 \times x - 80 = 4 \times y.$$

Rearranging we then need to solve the system of equations

$$\begin{cases} x+y & = & 100, \\ 2x-4y & = & 80. \end{cases}$$

Multiplying the first equation by 4,

$$4x+4y=400,$$

so if we add the two equations together we have

$$6x=480$$

and thus solving for x,

$$x=80.$$

Substituting into the first equation,

$$80+y=100$$

so solving for y,

$$y=100-80=20.$$

Hence there are 80 birds and 20 cats.

Problem 3.24 **How many gallons of a 25% alcohol solution must be mixed with a 50% alcohol solution to make 30 gallons of a 40% alcohol solution?**

Answer

12 gallons 25% solution, 18 gallons 50% solution

Solution 1

To have 30 gallons of 40% alcohol solution, we need

$$40\% \times 30 = 0.40 \times 30 = 12$$

gallons of alcohol. If we assume all 30 gallons are from the 25% alcohol solution, there would only be

$$25\% \times 30 = 0.25 \times 30 = 7.5$$

gallons of alcohol,
$$12 - 7.5 = 4.5$$

less than the actual amount. Therefore we must use the 50% alcohol solution. Each gallon of the 50% solution contains
$$50\% - 25\% = 25\%$$

more alcohol than the 25% solution. Therefore,
$$4.5 \div 25\% = 4.5 \div 0.25 = 18$$

gallons are the 50% alcohol solution. The remaining
$$30 - 18 = 12$$

gallons are from the 25% alcohol solution.

Solution 2

(Algebra) Let x be the number of gallons of 25% alcohol solution and y be the number of gallons of 50% alcohol solution. There are 30 gallons in total so
$$x + y = 30.$$

We want a 40% alcohol solution, so in total we need
$$40\% \times 30 = 0.40 \times 30 = 12$$

gallons of alcohol. Therefore
$$25\% \times x + 50\% \times y = 12.$$

Converting percents to decimals we get the system of equations
$$\begin{cases} x + y & = & 30, \\ 0.25x + 0.50y & = & 12. \end{cases}$$

Multiplying the second equation by 4 to clear the decimals we have
$$x + 2y = 48.$$

Subtracting this from the first equation, we have
$$y = 18.$$

Substituting back into the first equation,

$$x + 18 = 30$$

so we have

$$x = 30 - 18 = 12.$$

Thus we need 12 gallons of the 25% solution and 18 gallons of the 50% solution.

Problem 3.25 **A group of** 68 **people rent** 24 **motorcycles of two kinds at a race-track. The first kind has a capacity of** 2 **people and costs** $40 **per motorcycle. The second has a capacity of** 3 **people and costs** $30 **per motorcycle. The** 68 **people exactly fill all vehicles. What is the total cost in renting the** 24 **motorcycles?**

Answer

$760

Solution 1

If all 24 motorcycles are the first kind with a capacity of 2 people, the motorcycles will fit

$$24 \times 2 = 48$$

people, meaning that

$$68 - 48 = 20$$

people will left out. The second kind of motorcycle can hold

$$3 - 2 = 1$$

extra person, so if we switch 20 of the motorcycles to the second kind everyone will fit. Hence we have

$$24 - 20 = 4$$

motorcycles of the first kind and 20 of the second. Since the first kind costs $40 per motorcycle and the second kind costs $30, the total cost is

$$4 \times 40 + 20 \times 30 = 760$$

dollars to rent the 24 motorcycles.

Solution 2

(Algebra) Let x be the number of motorcycles of the first kind and y the number of the second kind that are rented. 24 motorcycles are rented in total, so

$$x + y = 24.$$

The 68 people all fit in these 24 motorcycles, with each motorcycle holding 2 or 3 people, so

$$2 \times x + 3 \times y = 68.$$

This gives the system of equations

$$\begin{cases} x + y & = & 24, \\ 2x + 3y & = & 68. \end{cases}$$

Doubling the first equation we have

$$2x + 2y = 48,$$

so subtracting this from the second equation we have

$$y = 20,$$

the number of motorcycles of the second kind. To find the number of the first kind, we substitute $y = 20$ back into the first equation to get

$$x + 20 = 24$$

so

$$x = 4,$$

the number of motorcycles of the first kind. These motorcycles each cost either \$40 or \$30 to rent. Therefore, the total cost is

$$4 \times 40 + 20 \times 30 = 760$$

to rent the motorcycles.

Problem 3.26 **There are many ducks and sheep in a farm. If we count the heads, there are total of 80 heads. If we count the legs, there are 56 more legs from sheep than from ducks. How many ducks and how many sheep are there in the farm.**

Answer

44 ducks, 36 sheep

Solution 1

If all 80 animals are sheep, there are

$$80 \times 4 = 320$$

sheep legs and 0 duck legs, so there are 320 more sheep legs than duck legs. If we replace one sheep with one duck we remove 4 sheep legs and add 2 duck legs, so the difference between the number of sheep legs and duck legs reduces by

$$4 + 2 = 6.$$

The difference we want is 56, which is

$$320 - 56 = 264$$

less than if we have all sheep. Thus we need to replace

$$264 \div 6 = 44$$

sheep with ducks. Hence there are

$$80 - 44 = 36$$

sheep and 44 ducks.

Solution 2

(Algebra) Let x be the number of ducks and y the number of sheep. Every animal has one head, so

$$x + y = 80.$$

Each duck has 2 legs and each sheep has 4 legs, so as there are 56 more sheep legs we have

$$2 \times x + 56 = 4 \times y.$$

Rearranging we have the system of equations

$$\begin{cases} x + y & = & 80, \\ 2x - 4y & = & -56. \end{cases}$$

Multiplying the first equation by 4 gives us

$$4x + 4y = 320.$$

Adding this to the second equation we have

$$6x = 264$$

so

$$x = \frac{264}{6} = 44.$$

Substituting,

$$44 + y = 80$$

so

$$y = 36.$$

Hence there are 44 ducks and 36 sheep.

Problem 3.27   **Bella goes shopping at the marketplace for shawls and belts. The shawls she likes each cost \$12. The belts she likes each cost \$14. Bella has exactly enough money to buy a certain number of shawls. If she buys belts instead, she has exactly enough money to buy 3 fewer belts. How much money did Bella bring with her to the market?**

Answer

   \$252

Solution 1

Bella can buy 3 more shawls than belts. These 3 shawls cost a total of

$$3 \times 12 = 36$$

dollars in total. Every shawl is

$$14 - 12 = 2$$

dollars cheaper than a belt. Hence if Bella buys

$$36 \div 2 = 18$$

shawls instead of 18 belts, she will have enough leftover money to buy 3 extra shawls. Hence Bella has exactly enough money to buy 18 belts, which is

$$18 \times 14 = 252$$

in total.

Solution 2

(Algebra) Let x be the number of belts Bella can buy. Then the number of shawls Bella can afford is $x + 3$. Buying x belts costs Bella

$$14 \times x$$

dollars, while buying $x + 3$ shawls costs

$$12 \times (x + 3).$$

Since Bella would spend all her money with either purchase, the two expressions are equal:

$$14x = 12 \times (x + 3).$$

Distributing and combining like terms we have

$$2x = 36$$

so dividing by 2 we have

$$x = \frac{36}{2} = 18.$$

Thus Bella has enough money to buy 18 belts. Hence she has

$$18 \times 14 = 252$$

dollars in total.

Problem 3.28 **Debbie is mixing orange juice concentrate for her restaurant. The first juice concentrate is 64% real orange juice. The second is only 48% real orange juice. How many ounces of 48% real orange juice should she use to make 1600 ounces of 58% real juice?**

Answer

60 ounces

Solution 1

If Debbie use 1600 ounces are from the first kind of orange juice, there would be

$$64\% \times 1600 = 0.64 \times 1600 = 1024$$

ounces of orange juice. She wants the juice to be 58% real orange juice, so she actually needs

$$58\% \times 1600 = 928$$

ounces of real orange juice. Hence she needs

$$1024 - 928 = 96$$

less ounces. Each ounce of the second kind contains

$$64\% - 48\% = 16\%$$

less orange juice than the first kind. Debbie needs 96 less ounces, so she needs to replace

$$96 \div 16\% = 96 \div 0.16 = 600$$

ounces of the first kind with the second kind. Hence Debbie should use 600 ounces of the 48% real orange juice.

Solution 2

(Algebra) Assume there are x ounces from the first kind and y ounces from the second kind. There are 1600 ounces in total, so

$$x + y = 1600.$$

In total we want a 58% juice mixture, so we need

$$58\% \times 1600 = 928$$

ounces of pure orange juice. Since the pure juice comes from either the first juice or the second,

$$64\% \times x + 48\% \times y = 928.$$

This gives the system of equations

$$\begin{cases} x + y & = & 1600, \\ 0.64x + 0.48y & = & 928. \end{cases}$$

Solving the first equation for x we get

$$x = 160 - y.$$

We can then plug this in the second equation to get

$$0.64 \times (1600 - y) + 0.48y = 928,$$

after distributing we have

$$1024 - 0.64y + 0.48y = 928$$

so by combining like terms we have

$$-0.16y = -96.$$

Therefore we can solve for y to get

$$y = \frac{-96}{-0.16} = \frac{96}{0.16} = 600,$$

the number of ounces of the second kind of orange juice.

Problem 3.29 **The Math Club collected donations from 40 people who live in either City A or B. Each person from City A contributed $5, and each person from City B contributed $8. In total, $5 more was collected from City A than from City B. How many people are there in each city?**

Answer

25 in City A, 15 in City B

Solution 1

If all the people were from City A, the Math Club would have collected $0 from City B and

$$40 \times 5 = 200$$

from City A. This is a difference of $200 dollars. Every person who is from City B instead of City A changes this difference by

$$5 + 8 = 13$$

because there is $5 less from City A and $8 more from City B. If we know that $5 more was collected from City A, which is

$$200 - 5 = 195$$

less the amount if all the people were from City A, we need to swap

$$195 \div 13 = 15$$

people from City A to City B. Hence there are

$$40 - 15 = 25$$

people from City A and 15 from City B.

Solution 2

(Algebra) Let x be the number of people from City A and y the number from City B. There are 40 donations in total, so

$$x + y = 40.$$

The club raised $5 more on the $5 donations from City A than on the $8 donations from City B, so

$$5 \times x = 8 \times y + 5.$$

Rearranging we have the system of equations

$$\begin{cases} x + y &= 40, \\ 5x - 8y &= 5. \end{cases}$$

Multiplying the first equation by 5 we get

$$5x + 5y = 200$$

and subtracting the second equation from this,

$$13y = 195.$$

Solving for y,

$$y = \frac{195}{13} = 15,$$

so substituting into the first equation,

$$x + 15 = 40$$

and

$$x = 40 - 15 = 25.$$

Hence there are 25 people from City A and 15 people from City B.

Problem 3.30 **A butcher has some hamburger meat that is** 4% **fat and some hamburger meat that is** 20% **fat. How much of each type will he need to make** 120 **pounds of hamburger meat which is** 10% **fat?**

Answer

4%: 75 pounds, 20%: 45 pounds

Solution 1

If the Butcher made all 120 pounds using the 4% fat meat, it would only have

$$4\% \times 120 = 0.04 \times 120 = 4.8$$

pounds of fat. They want a 10% fat mixture, so the actual amount of fat needed is

$$10\% \times 120 = 12$$

pounds. Hence, the butcher needs

$$12 - 4.8 = 7.2$$

extra pounds of fat. Each pound of the 20% fat meat contains

$$20\% - 4\% = 16\%$$

more fat, and thus if the butcher switches

$$7.2 \div 16\% = 7.2 \div 0.16 = 45$$

pounds of meat to the 20% fat meat he will have the correct amount of fat. The remaining

$$120 - 45 = 75$$

pounds are from the 4% fat hamburger meat.

Solution 2

(Algebra) Assume there are x pounds from the 4% and y pounds from the 20% fat meats. The butcher needs 120 pounds in total, so

$$x + y = 120.$$

They want a 10% fat mixture, so the amount of fat needed is

$$10\% \times 120 = 12$$

pounds. This comes from the 4% and 20% fat meats, so

$$4\% \times x + 20\% \times y = 12.$$

This gives the system of equations

$$\begin{cases} x + y & = & 120, \\ 0.04x + 0.2y & = & 12. \end{cases}$$

Note that

$$4\% = 0.04 = \frac{1}{25},$$

so multiplying the second equation by 25 we have

$$x + 5y = 300.$$

Subtracting the first equation from this,

$$4y = 180,$$

so

$$y = \frac{180}{4} = 45.$$

Substituting into the first equation,

$$x + 45 = 120,$$

so

$$x = 120 - 45 = 75.$$

Therefore, the butcher should use 75 pounds of the 4% fat hamburger meat and 45 pounds of the 20% fat meat.

Problem 3.31 **Hank has a bottle of diluted syrup that is** 60% **maple syrup and a bottle of pure syrup that is** 100% **maple syrup in his restaurant. How many ounces of each should he mix in order to make** 100 **ounces of 85% maple syrup?**

Answer

60%: 37.5, 100%: 62.5

Solution 1

Since 100 ounces of 85% maple syrup contains

$$85\% \times 100 = 0.85 \times 100 = 85$$

ounces of pure maple syrup, ff Hank starts with 100 ounces of the pure syrup, he will have

$$100 - 85 = 15$$

more ounces of pure syrup than he wants. Each ounce of the diluted 60% syrup contains

$$100\% - 60\% = 40\%$$

less syrup than the pure syrup, so if Hank replaces

$$15 \div 40\% = 15 \div 0.40 = 37.5$$

ounces of the pure syrup with the diluted syrup he will have the correct 85% mixture. Hence he needs

$$100 - 37.5 = 62.5$$

ounces of the pure syrup and 37.5 ounces of the diluted syrup.

Solution 2

(Algebra) Assume there are x ounces of the diluted syrup and y ounces of the pure syrup. Hank needs 100 ounces is total, so

$$x + y = 100.$$

He wants a mixture that contains 85% maple syrup, so he needs a total of

$$85\% \times 100 = 0.85 \times 100 = 85$$

ounces of maple syrup. Since the diluted syrup is 60% maple syrup and the pure syrup is 100% maple syrup,

$$60\% \times x + 100\% \times y = 85.$$

Hence we have the system of equations

$$\begin{cases} x+y &=& 100, \\ 0.6x+y &=& 85. \end{cases}$$

Subtracting the second equation the first we get

$$0.4x = 15$$

so we can solve for x to get

$$x = \frac{15}{0.4} = 37.5.$$

Using the first equation we then have

$$37.5 + y = 100$$

so

$$y = 100 - 37.5 = 62.5.$$

Therefore Hank should use 37.5 ounces of the diluted maple syrup and 62.5 ounces of the pure maple syrup.

Problem 3.32   **How many gallons of 60% antifreeze should be mixed with** 40% **antifreeze to make** 80 **gallons of** 45% **antifreeze?**

Answer

20

Solution 1

To make 80 gallons of 45% antifreeze, we need

$$45\% \times 80 = 36$$

gallons of pure antifreeze. If have 80 gallons of the 60% antifreeze we would instead have

$$60\% \times 80 = 48$$

gallons, which is

$$48 - 36 = 12$$

more than needed. Replacing one gallon of the 60% antifreeze with 40% antifreeze is a

$$60\% - 40\% = 20\%$$

reduction of the antifreeze for that gallon. Thus if need to replace

$$12 \div 20\% = 12 \div 0.2 = 60$$

gallons of the 60% antifreeze with 40% to get the correct mixture. Hence we use

$$80 - 60 = 20$$

gallons of the 60% antifreeze in total.

Solution 2

(Algebra) Assume there are x gallons of the 60% antifreeze and y gallons of the 40%. We want 80 gallons in total, so

$$x + y = 80.$$

To achieve a 45% antifreeze mixture, we need

$$45\% \times 80 = 36$$

gallons of pure antifreeze. Since our two mixtures are 60% and 40% respectively,

$$45\% \times x + 60\% \times y = 36.$$

Combining and simplifying we have the system of equations

$$\begin{cases} x + y & = & 80, \\ 0.6x + 0.4y & = & 36. \end{cases}$$

Solving the first equation for y we have

$$y = 80 - x$$

and substituting this into the second equation we have

$$0.6x + 0.4 \times (80 - x) = 36.$$

Distributing we have

$$0.6x + 32 - 0.4x = 36,$$

so after combining like terms we get

$$0.2x = 4.$$

Hence solving for x,

$$x = \frac{4}{0.2} = 4 \times 5 = 20.$$

Thus we should use 20 gallons of the 60% antifreeze in our mixture.

Problem 3.33 **It requires either 45 small trucks or 36 big trucks to transport a batch of steel blocks. Given that each big truck can load 4 more tons of steel blocks than each small truck. How many tons of steel blocks are in a batch?**

Answer

720

Solution 1

A full batch can be carried by 36 big trucks. Since each big truck can load 4 more tons than each small truck, if 36 small trucks are used, there is still

$$36 \times 4 = 144$$

tons of steel blocks left over. However, we know that 45 small trucks in total can carry the full batch, so the

$$45 - 36 = 9$$

extra trucks must carry the extra 144 tons. Hence each small truck can carry

$$144 \div 9 = 16$$

tons of steel blocks. Finally, this means that one batch, which is carried by 45 small trucks, is a total of

$$45 \times 16 = 720$$

tons of steel blocks.

Solution 2

The algebra method. Let the capacity of a small truck be x tons, then the capacity of each big truck is $x+4$ tons. Since one batch is carried by 45 small trucks or by 36 big trucks, we have

$$45 \times x = 36 \times (x+4).$$

Distributing and combining like terms we have

$$9x = 144.$$

Hence we can solve for

$$x = \frac{144}{9} = 16.$$

Be careful, as x is just the capacity of one small truck. Since one batch is carried by 45 small trucks, one batch is

$$45 \times 16 = 720$$

tons of steel blocks.

Problem 3.34 **In a farm the total number of chickens and rabbits is** 100. **If the number of chicken feet is** 80 **more than the number of rabbit feet, how many chickens and rabbits are there respectively?**

Answer

80 chickens, 20 rabbits

Solution 1

If there are only chickens on the farm, there are 100 chickens and therefore

$$100 \times 2 = 200$$

chicken feet. Since there are no rabbits, there are 200 more chicken feet than rabbit feet. If we replace one chicken with a rabbit, we remove 2 chicken feet and add 4 rabbit feet, a total change in

$$2+4 = 6$$

of the difference between chicken and rabbit feet. Since we want this difference to be 80 and

$$200 - 80 = 120,$$

we need to replace

$$120 \div 6 = 20$$

chickens with rabbits. Therefore the farm has

$$100 - 20 = 80$$

chickens and 20 rabbits.

Solution 2

(Algebra) Let x be the number of chickens and y be the number of rabbits on the farm. There are 100 animals in total, so

$$x + y = 100.$$

Each chicken has two feet and each rabbit has 4 feet. There are 80 more chicken feet than rabbit feet on the farm, so

$$2 \times x = 4 \times y + 80,$$

which after rearranging gives us the system of equations

$$\begin{cases} x + y & = & 100, \\ 2x - 4y & = & 80. \end{cases}$$

Doubling the first equation we get

$$2x + 2y = 200.$$

Subtracting the second equation from this we get

$$6y = 120.$$

Therefore

$$y = \frac{120}{6} = 20.$$

Substituting we have

$$x + 20 = 100$$

so

$$x = 100 - 20 = 80.$$

Hence there are 80 chickens and 20 rabbits on the farm.

Problem 3.35 **Four basketballs and five volleyballs cost** 185 **dollars in total. If a basketball costs** 8 **dollars more than a volleyball, what is the cost of one basketball?**

Answer

25 dollars

Solution 1

There are a total of
$$4 + 5 = 9$$
balls bought in total. If we tried to save money and bought nine volleyballs instead of four basketballs and five volleyballs, then the price would be
$$4 \times 8 = 32$$
dollars cheaper because each volleyball is $8 less than a basketball. Therefore 9 volleyballs cost
$$185 - 32 = 153$$
dollars in total. Hence we can fine the price of one volleyball which is
$$153 \div 9 = 17$$
dollars. Since one basketball is $8 more, a single basketball costs
$$17 + 8 = 25$$
dollars.

Solution 2

(Algebra) Let x be the price of one volleyball. Since a basketball is $8 more expensive, it costs
$$x + 8$$
dollars. We know that four basketballs and five volleyballs cost $185, so
$$4 \times (x + 8) + 5 \times x = 185.$$

Distributing and combining like terms we have

$$9x = 163$$

so

$$x = \frac{163}{9} = 17,$$

the price of one volleyball. A basketball therefore costs

$$17 + 8 = 25$$

dollars.

Problem 3.36 A turtle has 4 legs and a crane has 2 legs. There are totally 100 heads of turtles and cranes, and there are 20 more crane legs than turtle legs. How many of each animal are there?

Answer

30 turtles, 70 cranes

Solution 1

There are 20 more crane legs than turtle legs. Each turtle has 4 legs, so if we assume

$$20 \div 4 = 5$$

additional turtles are added, then there are

$$100 + 5 = 105$$

heads altogether and the number of crane legs and turtle legs are the same. Since each turtle has twice as many legs as a crane, there must be half as many turtles. Since

$$105 \div 3 = 35$$

there are 35 turtles and

$$2 \times 35 = 70$$

cranes now. Removing the added 5 turtles, the original numbers are

$$35 - 5 = 30$$

turtles and 70 cranes.

Solution 2

(Algebra) Let x be the number of turtles and y the number of cranes. There are 100 heads in total, so

$$x + y = 100.$$

We know there are 20 more crane legs than turtle legs. Hence

$$4x + 20 = 2y.$$

After combining like terms we get the system of equations

$$\begin{cases} x + y & = & 100, \\ 4x - 2y & = & -20. \end{cases}$$

Multiplying the first equation by 2 we get

$$2x + 2y = 200,$$

and then adding this to the second equation we have

$$6x = 180.$$

Hence we can solve for x to get

$$x = \frac{180}{6} = 30$$

as the number of turtles. Substituting into the first equation in our system,

$$30 + y = 100$$

so

$$y = 100 - 30 = 70,$$

the number of cranes.

Problem 3.37 **On a good day, Chris the Squirrel picks 20 hazelnuts. On a rainy day he only picks 12 hazelnuts. During a few consecutive days he picked a total of 120 hazelnuts with an average of 15 per day. How many days were rainy?**

Answer

5

Solution 1

First note that since Chris picked 15 hazelnuts on average and we know that he picked 120 hazelnuts in total, there must have been

$$120 \div 15 = 8$$

days in total. If each of the 8 days were sunny, Chris would have picked

$$8 \times 20 = 160$$

total hazelnuts,

$$160 - 120 = 40$$

more than he actually did. On a rainy day Chris picks

$$20 - 12 = 8$$

fewer hazelnuts than on a sunny day. Therefore,

$$40 \div 8 = 5$$

of the 8 days must have been rainy.

Solution 2

(Algebra) Let x be the number of sunny days and y be the number of rainy days. Chris picks a total of 120 hazelnuts. Since he averages 15 per day, there are

$$120 \div 15 = 8$$

total days, so

$$x + y = 8.$$

Each sunny day Chris can pick 20 hazelnuts and each rainy day he picks 12 hazelnuts. Since he picks 120 in total,

$$20 \times x + 12 \times y = 120.$$

We thus need to solve the system of equations

$$\begin{cases} x + y & = & 8, \\ 20x + 12y & = & 120. \end{cases}$$

Multiplying the first equation by 20, we have

$$20x + 20y = 160.$$

Subtracting the second equation from this gives

$$8y = 40,$$

so we can solve for y and get

$$y = \frac{40}{8} = 5.$$

Hence 5 of the days are rainy.

Problem 3.38 **In a farm there are 6 times as many rabbits as chickens, and the total number of feet of the chickens and rabbits is 390. What is the number of each type of the animals?**

Answer

15 chickens, 90 rabbits

Solution 1

We know there are 6 rabbits for every chicken. This means we can group all the animals on the farm into groups of 7, with each group containing one chicken and six rabbits. Therefore each group has a total of

$$2 + 6 \times 4 = 26,$$

legs. Since there are 390 legs in total, there are

$$390 \div 26 = 15$$

groups of animals. Hence there are 15 chickens and

$$6 \times 15 = 90$$

rabbits on the farm.

Solution 2

(Algebra) Let x be the number of chickens on the farm. There are 6 times as many rabbits, so we know there are

$$6 \times x = 6x$$

rabbits on the farm. Each chicken has 2 legs and each rabbit has 4 legs. This gives us

$$2 \times x + 4 \times (6x) = 390.$$

Combining like terms, we get

$$26x = 390$$

so

$$x = \frac{390}{26} = 15,$$

the number of chickens on the farm. We lastly calculate there are

$$6 \times 15 = 90$$

rabbits.

Problem 3.39 **The capacity of a big container is** 4 **gallons, and that of a small container is** 2 **gallons.** 50 **containers are filled with water, and there are totally** 20 **more gallons of water in the big containers than the small containers. How many big and small containers are there respectively?**

Answer

20 big containers, 30 small containers

Solution 1

20 more gallons of water are stored in the big containers than in the small containers. Since each small container holds 2 gallons of water, if we add 10 more small containers full of water, then the amount of water stored in the big containers is the same as in the small containers. Each big container contains twice as much water as a small container, so if an equal amount of water is stored in both there must be twice as many small containers. Of the

$$50 + 10 = 60$$

containers in this case,

$$60 \div 3 = 20$$

must be big and

$$2 \times 20 = 40$$

must be small. Recalling we added 10 extra small containers, there are actually

$$40 - 10 = 30$$

small containers and 20 big containers.

Solution 2

(Algebra) Let x be the number of big containers and y be the number of small containers. There are 50 containers in total, so

$$x + y = 50.$$

20 more gallons are stored in the big containers, so as each big container holds 4 gallons and each small holds 2 gallons,

$$4 \times x = 2 \times y + 20.$$

This gives us the system of equations

$$\begin{cases} x + y & = & 50, \\ 4x - 2y & = & 20. \end{cases}$$

Multiplying the first equation by 2,

$$2x + 2y = 100.$$

Adding this with the second equation we have

$$6x = 120.$$

This allows us to solve for x to get

$$x = \frac{120}{6} = 20,$$

the number of big containers. Substituting into the first equation,

$$20 + y = 50,$$

so there are

$$y = 50 - 20 = 30$$

small containers.

Problem 3.40   **Morgan needed 70 sticks for a project at school. Each stick is either 3 inches or 5 inches and the total length of all the sticks combined is 270 inches. How many 3 inch sticks and 5 inch sticks are there?**

Answer

40 three inch, 30 five inch sticks.

Solution 1

If all the sticks were 3 inch sticks, the total length of the 70 sticks would be

$$70 \times 3 = 210$$

inches. Since the actual length is

$$270 - 210 = 60$$

inches longer, some of the 3 inch sticks need to be replaced with 5 inch sticks. Each 5 inch stick is

$$5 - 3 = 2$$

inches longer. Since the actual length is 60 more than the length of all 3 inch sticks, we must replace

$$60 \div 2 = 30$$

3 inch sticks with 5 inch sticks. Hence there are

$$70 - 30 = 40$$

three inch sticks and 30 five inch sticks.

Solution 2

(Algebra) Let x be the number of 3 inch sticks and y be the number of 5 inch sticks. There are 70 sticks in total, so

$$x + y = 70.$$

The total combined length of the 3 and 5 inch sticks is 270 so we also have

$$3 \times x + 5 \times y = 260.$$

This gives the system of equations

$$\begin{cases} x+y &= 70, \\ 3x+5y &= 270. \end{cases}$$

Multiplying the first equation by 3 we get

$$3x+3y=210$$

and subtracting this from the second equation we have

$$2y=60.$$

Hence solving for x we have

$$y=30,$$

the number of 5 inch sticks. Lastly, substituting back into the first equation,

$$x+30=70$$

so

$$x=70-30=40,$$

the number of 3 inch sticks.

Problem 3.41 **Some chickens and rabbits have a total of 100 feet. If each chicken was exchanged for a rabbit, and each rabbit was exchanged for a chicken, there would be a total of 86 feet. How many chickens are there? How many rabbits?**

Answer

12 chickens, 19 rabbits

Solution 1

We know that there are 100 feet total. We first find out how many animals there are in total. Pretend that for every chicken on the farm, we pair it up with a new rabbit, and for every rabbit on the farm, we pair it up with a new chicken. Note that this adds a total of 86 feet. Hence there are a total of

$$100+86=186$$

feet. Each chicken and rabbit pair has a combined total of

$$2 + 4 = 6$$

feet, so there must be

$$186 \div 6 = 31$$

chicken and rabbit pairs. As every original animal is in exactly one pair, this means there are 31 animals on the farm.

If all 31 animals were chickens, there would be a total of

$$31 \times 2 = 62$$

feet, which is

$$100 - 62 = 38$$

less than the true amount. As each rabbit has

$$4 - 2 = 2$$

extra feet, if we change

$$38 \div 2 = 19$$

chickens to rabbits we will have the correct number of feet. Hence there are

$$31 - 19 = 12$$

chickens and 19 rabbits.

Solution 2

(Algebra) Let x be the number of chickens and y be the number of rabbits. There are 100 feet in total, so as each chicken has 2 feet and each rabbit has 4,

$$2 \times x + 4 \times y = 100.$$

We also know that swapping all the animals we have 86 feet, so

$$4 \times x + 2 \times y = 86.$$

This gives the system of equations

$$\begin{cases} 2x + 4y &= 100, \\ 4x + 2y &= 86. \end{cases}$$

Doubling the first equation gives us

$$4x + 8y = 200.$$

We can then subtract the second equation from this to get

$$6y = 114,$$

and dividing by 6 we have

$$y = \frac{114}{6} = 19.$$

Substituting into the first equation,

$$2x + 4 \times 19 = 100$$

so combining like terms we have

$$2x = 24.$$

Hence

$$x = \frac{24}{2} = 12$$

so there are 12 chickens and 19 rabbits.

Problem 3.42 100 **mice eat** 100 **cakes. If each big mouse eats** 3 **cakes, and** 3 **baby mice eat** 1 **cake, how many big mice and baby mice are there?**

Answer

25 big mice, 75 baby mice

Solution 1

We can group them and then solve the problem. Let's group one big mouse and three baby mice into a group, then in this group, there are four mice and they will eat

$$3 + 1 = 4$$

cakes. Since there are 100 cakes in total, there must be

$$100 \div 4 = 25$$

groups. In each group, there is only one big mouse, so the total number of big mice is

$$25 \times 1 = 25.$$

In each group, there are 3 baby mice, so there are

$$25 \times 3 = 75$$

small mice in total.

Solution 2

Since 3 baby mice eat 1 cake, one baby mouse only eats

$$1 \div 3 = \frac{1}{3}$$

of a cake. Hence a big mouse eats

$$3 - \frac{1}{3} = \frac{9}{3} - \frac{1}{3} = \frac{8}{3}$$

more cakes than a baby mouse. If all 100 mice are big, then they would eat

$$100 \times 3 = 300$$

cakes. In fact, there are only 100 cakes, which is

$$300 - 100 = 200$$

less than if there were all big mice. Since a big mouse eats $\frac{8}{3}$ more than a baby mouse, if we change

$$200 \div \frac{8}{3} = 75$$

big mice to baby mice the mice will eat the correct number of cakes. Therefore there are 75 baby mice and

$$100 - 75 = 25$$

big mice.

Solution 3

(Algebra) Let x be the number of big and y be the number of baby mice. There are 100 mice in total, so

$$x + y = 100.$$

3 baby mice eat 1 cake, so one baby mouse eats

$$1 \div 3 = \frac{1}{3}$$

of a cake. Since we also know a big mouse eats 3 cakes,

$$3 \times x + \frac{1}{3} \times y = 100.$$

Therefore we have the system of equations

$$\begin{cases} x + y & = & 100, \\ 3x + y/3 & = & 100. \end{cases}$$

Multiplying the second equation by 3 we get

$$9x + y = 300.$$

Subtracting the first equation from this gives us

$$8x = 200,$$

so we solve for x,

$$x = \frac{200}{8} = 25.$$

Plugging back into the first equation,

$$25 + y = 100,$$

so

$$y = 100 - 25 = 75.$$

Thus there are 25 big mice and 75 baby mice.

Problem 3.43 **A spider has 8 legs. A firefly has 6 legs and 2 pairs of wings. A cicada has 6 legs and 1 pair of wings. There are a total of 16 bugs of the three types. There are 110 legs in total. There are 14 pairs of wings in total. How many of each kind of bug are there?**

Answer

 7 spiders, 5 fireflies, 4 cicadas

Solution 1

 If we first assume all 16 bugs are fireflies, then there will be

$$16 \times 6 = 96$$

legs. There are in fact 110 legs, which is

$$110 - 96 = 14$$

more than if we all bugs were fireflies. Each cicada has the same number of legs but each spider has

$$8 - 6 = 2$$

extra legs, so there must be

$$14 \div 2 = 7$$

spiders in total. The number of fireflies and cicadas is thus

$$16 - 7 = 9.$$

Again, assume these 9 bugs are all fireflies. There will be

$$9 \times 2 = 18$$

pairs of wings. In fact, we only have 14 pairs, a difference of

$$18 - 14 = 4$$

pairs of wings. Each firefly has

$$2 - 1 = 1$$

more pair of wings than a cicada, so there must be

$$4 \div 1 = 4$$

cicadas. Hence the number of fireflies is

$$9 - 4 = 5.$$

To summarize there are 7 spiders, 5 fireflies, and 4 cicadas.

Solution 2

(Algebra) Let x be the number of spiders, y be the number of fireflies, and z be the number of cicadas. First we know the total number of bugs is 16, so

$$x + y + z = 16.$$

Looking at the number of legs, spiders have 8, while fireflies and cicadas have 6, so

$$8 \times x + 6 \times y + 6 \times z = 110,$$

the total number of legs. Finally, there are 14 pairs of wings, so counting the number of pairs for each type of bug we have

$$0 \times x + 2 \times y + 1 \times z = 14.$$

This gives the three variable three equation system of equations:

$$\begin{cases} x + y + z & = & 16, \\ 8x + 6y + 6z & = & 110, \\ 2y + z & = & 14. \end{cases}$$

Multiplying the first equation by 6,

$$6x + 6y + 6z = 96,$$

so subtracting this from the second equation gives us that

$$2x = 14$$

so

$$x = \frac{14}{2} = 7,$$

the number of spiders. Substituting this into the second equation,

$$8 \times 7 + 6y + 6z = 110,$$

so simplifying we need to solve two variable two equation system

$$\begin{cases} 6y + 6z & = & 54, \\ 2y + z & = & 14. \end{cases}$$

Dividing the first equation by 6,

$$y + z = 9.$$

Subtracting this from the second equation

$$y = 5,$$

the number of fireflies. Substituting we have

$$5 + z = 9$$

so

$$z = 9 - 5 = 4$$

the number of cicadas.

Problem 3.44   **The school purchases 3 different sizes of projectors, total of 47. The large size costs \$700, the medium costs \$300, and the small costs \$200. The total cost of the projectors is \$21200, and there are twice as many medium projectors than the small. How many large projectors does the school purchase?**

Answer

20

Solution 1

If the school bought only large projectors, they would spend a total of

$$47 \times 700 = 32900$$

dollars, which is

$$32900 - 21200 = 11700$$

dollars over their budget of 21200. Since there are twice as many medium projectors as small projectors, we can view them as coming in groups of 3, with 1 small and 2 medium projectors. One such group costs a total of

$$1 \times 200 + 2 \times 300 = 800$$

dollars, which is

$$3 \times 700 - 800 = 1300$$

dollars cheaper than a group of 3 large projectors. Hence if the school switchs

$$11700 \div 1300 = 9$$

groups of 3 large projectors to groups of 1 small, 2 medium projectors the school will spend the correct amount of money. Thus, the school buys

$$47 - 3 \times 9 = 47 - 27 = 20$$

large projectors.

Solution 2

(Algebra) Let x be the number of small projectors, so we know the number of medium projectors is
$$2 \times x = 2x.$$

Let y be the number of large projectors. There are 47 projectors bought in total, so
$$x + 2x + y = 47,$$

for a total cost of $21200, so

$$200 \times x + 300 \times (2x) + 700 \times y = 21200.$$

Simplifying these two equations we get the system of equations

$$\begin{cases} 3x + y & = & 47, \\ 800x + 700y & = & 21200. \end{cases}$$

Dividing the second equation by 100 we get a simplified version of

$$8x + 7y = 212.$$

Solving the first equation for y,

$$y = 47 - 3x.$$

Plugging this into the other equation we get

$$8x + 7 \times (47 - 3x) = 212,$$

or
$$8x + 329 - 21x = 212$$

after distributing. Hence we get

$$-13x = -117$$

after combining like terms and hence

$$x = 9.$$

Therefore,

$$y = 47 - 3 \times 9 = 20.$$

Thus the school purchases 20 large projectors.

Problem 3.45 **Tony's mom took out \$380 from the bank. There are \$2, \$5, and \$10 bills and total of 80. The number of \$5 bills and \$10 bills are the same. How many bills of each type are there?**

Answer

40 \$2, 20 \$5, 20 \$10 bills

Solution 1

We know there are equal numbers of \$5 and \$10 bills, so pairs these bills up for a total of

$$5 + 10 = 15$$

dollars per pair. If Tony's mom has only \$5 and \$10 bills, there will be

$$80 \div 2 = 40$$

pairs for a total of

$$40 \times 15 = 600$$

dollars, which is

$$600 - 380 = 220$$

more than she actually has. A pair of \$2 bills is worth

$$15 - 2 \times 2 = 11$$

dollars less than a \$5 and \$10 bill pair, so there must be

$$220 \div 11 = 20$$

pairs of \$2 bills. Hence there are

$$40 - 20 = 20$$

$5 and $10 bill pairs. Tony's mom therefore has

$$20 \times 2 = 40$$

$2 bills, 20 $5 bills, and 20 $10 bills.

Solution 2

(Algebra) Let the number of $2 bills be x. Let y be the number of $5 bills, so there are y $10 bills as well. There are 80 bills in total, so

$$x + y + y = 80.$$

The $2, $5, $10 bills in total are worth $380 in total, so we also have

$$2 \times x + 5 \times y + 10 \times y = 380.$$

This gives the system of equations

$$\begin{cases} x + 2y & = & 80, \\ 2x + 15y & = & 380. \end{cases}$$

Doubling the first equation

$$2x + 4y = 160,$$

and subtracting this from the second equation we have

$$11y = 220$$

and therefore

$$y = 20.$$

Plugging this into the first equation,

$$x + 2 \times 20 = 80,$$

so

$$x = 80 - 40 = 40.$$

Hence there are 40 $2 bills, 20 $5 bills, and 20 $10 bills.

Problem 3.46 **Cindy collects 20 insects for her biology class, all of which are spiders, dragonflies, and cicadas. (Note that a spider has 8 legs and no wings, a dragonfly has 6 legs and 4 wings, and a cicada has 6 legs and 2 wings.) She counts 138 legs and 36 wings altogether. How many insects are there in each kind?**

Answer

9 spiders, 7 dragonflies, and 4 cicadas

Solution 1

Note first that dragonflies and cicadas each have 6 legs. If all 20 insects are spiders, there will be a total of

$$20 \times 8 = 160$$

legs, which is

$$160 - 138 = 22$$

more than the total amount. Each spider has

$$8 - 6 = 2$$

extra legs compared to dragonflies and cicadas, so

$$22 \div 2 = 11$$

of the bugs must be either dragonflies or cicadas, so

$$20 - 11 = 9$$

are spiders. For the non-spiders, there are 36 wings in total. If all 9 of these bugs are dragonflies, there would be

$$11 \times 4 = 44$$

wings, which is

$$44 - 36 = 8$$

more than the actual amount. Each dragonfly has

$$4 - 2 = 2$$

more wings than a cicada, so

$$8 \div 2 = 4$$

bugs must be cicadas. The remaining

$$11 - 4 = 7$$

are dragonflies.

Solution 2

(Algebra) Let x be the number of spiders, y be the number of dragon-flies, and z be the number of cicadas. Counting the total number of bugs we have

$$x + y + z = 20.$$

Counting the total number of legs we have

$$8 \times x + 6 \times y + 6 \times z = 138.$$

Lastly, counting the total number of wings we have

$$0 \times x + 4 \times y + 2 \times z = 36.$$

Combining we have the system of equations

$$\begin{cases} x + y + z & = & 20, \\ 8x + 6y + 6z & = & 138, \\ 4y + 2z & = & 36. \end{cases}$$

Multiplying the first equation by 6 we have

$$6x + 6y + 6z = 120,$$

and subtracting this from the second gives

$$2x = 18$$

and hence

$$x = \frac{18}{2} = 9.$$

Thus we can rewrite the second equation as

$$8 \times 9 + 6 \times y + 6 \times z = 138$$

or

$$6y + 6z = 66.$$

Multiplying the third equation by 3,

$$12y + 6z = 108.$$

Hence subtracting these two equation we have

$$6y = 108 - 66 = 42,$$

and

$$y = \frac{42}{6} = 7.$$

Lastly, $x = 9, y = 7$ into the first equation,

$$9 + 7 + z = 20$$

so

$$z = 20 - 9 - 7 = 4.$$

There are 9 spiders, 7 dragonflies, and 4 cicadas.

Problem 3.47 **A candy shop sold three flavors of candies, cherry, strawberry, and watermelon, in the morning. The prices are $20/kg, $25/kg, and $30/kg, respectively. The shop sold a total of 100 kg and received $2570. It is known that the total sale of cherry and watermelon flavor candies combined is $1970. How many kilograms of watermelon flavor candies were sold?**

Answer

45

Solution 1

We know the total sale was $2570. Since $1970 of this was for cherry and watermelon, the remaining

$$2570 - 1970 = 600$$

dollars was due to strawberry. Hence

$$600 \div 25 = 24$$

kilograms of strawberry candy was sold. Thus the other

$$100 - 24 = 76$$

kilograms were cherry and watermelon. If thse 76 kg were all cherry, the total sales would be

$$76 \times 20 = 1520$$

dollars, which is

$$1970 - 1520 = 450$$

less than the actual amount. As each kg of watermelon is

$$30 - 20 = 10$$

dollars per kg more expensive, there must have been

$$450 \div 10 = 45$$

kg of watermelon flavor candies sold.

Solution 2

Let x be the amount of cherry candy sold in kg, y the amount of strawberry candy, and z the amount of watermelon candy. There is 100 kg sold in total, so

$$x + y + z = 100.$$

We then know cherry costs \$20 per kg, strawberry \$25 per kg, and watermelon \$30 per kg, so we have that

$$20 \times x + 25 \times y + 30 \times z = 2570,$$

the total amount sold in dollars. We also know \$1970 of this was just cherry and watermelon, so

$$20 \times x + 30 \times z = 1970.$$

This gives the system of equations

$$\begin{cases} x + y + z & = & 100, \\ 20x + 25y + 30z & = & 2570, \\ 20x + 30z & = & 1970. \end{cases}$$

Noticing the similarities between the second and third equations, we see that if we subtract the third from the second we get

$$25y = 2570 - 1970 = 600$$

so we can solve for

$$y = \frac{600}{25} = 24.$$

Plugging this into the first equation,

$$x + 24 + z = 100$$

so simplifying and multiplying by 20 we get

$$20x + 20z = 1520.$$

Subtracting this from the third equation we have

$$10z = 1970 - 1520 = 400.$$

Hence,

$$z = \frac{400}{10} = 40,$$

the number of kg of watermelon candy sold, as needed.

Problem 3.48 **A crab has 10 legs. A mantis has 6 legs and 1 pair of wings. A dragonfly has 6 legs and 2 pairs of wings. There are a total of 37 of the three types. There are 250 legs in total. There are 52 pairs of wings in total. How many of each kind are there?**

Answer

7 crabs, 8 mantises, 22 dragonflies

Solution 1

If all 37 animals are mantises, there will be a total of

$$37 \times 6 = 222$$

legs, which is

$$250 - 222 = 28$$

legs more than the actual amount. Each dragonfly also has 6 legs, but each crab has

$$10 - 6 = 4$$

extra legs. Hence there must be

$$28 \div 4 = 7$$

crabs to give us the 28 extra legs. The remaining

$$37 - 7 = 30$$

are either mantises or dragonflies. They each have the same number of legs, so we need to look at the pairs of wings. If we again assume all 30 remaining are mantises, there will be a total of

$$30 \times 1 = 30$$

pairs of wings, which is

$$52 - 30 = 22$$

less than the actual amount. Each dragonfly has

$$2 - 1 = 1$$

extra pair of wings, so there must be 22 dragonflies in total. Lastly, the remaining

$$30 - 22 = 8$$

are mantises. In all there are 7 crabs, 8 mantises, and 22 dragonflies.

Solution 2

(Algebra) Let x be the number of crabs, and y be the number of mantises, and z be the number of dragonflies. By the tota number of animals we have

$$x + y + z = 37.$$

Using the number of legs,

$$10 \times x + 6 \times y + 6 \times z = 250.$$

Finally, using the number of pairs of wings,

$$0 \times x + 1 \times y + 2 \times z = 52.$$

This gives us the system of equations

$$\begin{cases} x + y + z & = & 37, \\ 10x + 6y + 6z & = & 250, \\ y + 2z & = & 52. \end{cases}$$

Multiplying the first equation by 6,

$$6x + 6y + 6z = 222,$$

so after subtracting this from the second equation we have

$$4x = 28$$

and therefore,

$$x = \frac{28}{4} = 7.$$

Plugging this into the second equation,

$$10 \times 7 + 6y + 6z = 250$$

so

$$6y + 6z = 180.$$

Dividing this by 6 we have

$$y + z = 30.$$

Subtracting this from the third equation,

$$z = 22,$$

and therefore

$$y + 22 = 30$$

so

$$y = 30 - 22 = 8.$$

Hence, $x = 7$, $y = 8$, and $z = 22$, so there are 7 crabs, 8 mantises, and 22 dragonflies.

Problem 3.49 **100 monks eat 100 steamed buns. If each senior monk eats 4 steamed buns, and 4 junior monks eat 1 steamed bun, how many senior monks and junior monks are there?**

Answer

20 senior and 80 junior monks

Solution 1

Grouping one senior monk with four junior monks, we have a group of 5 monks that eats

$$4 + 1 = 5$$

steamed buns in total. Since there are 100 cakes in total, there must be

$$100 \div 5 = 20$$

such groups. In each group, there is only one senior monk, so there are

$$20 \times 1 = 20$$

senior monks in total. Similarly, in each group, there are 4 junior monks, so there are

$$20 \times 4 = 80$$

junior monks in total.

Solution 2

Since 4 junior monks eat 1 steamed bun, one junior monk eats

$$1 \div 4 = 0.25$$

of a steamed bun. As a senior monk eats 4 steamed buns, a senior monk eats

$$4 - 0.25 = 3.75$$

more buns than a junior monk. If all 100 monks were senior, they would eat a total of

$$100 \times 4 = 400$$

buns. However, there are only 100 steamed buns, so this is

$$400 - 100 = 300$$

buns too many. We know a senior monk eats 3.75 more buns than a junior monk, so if we change

$$300 \div 3.75 = 80$$

senior monks to junior monks, the monks will eat the correct number of steamed buns. Therefore there are 80 junior monks and

$$100 - 80 = 20$$

senior monks.

Solution 3

(Algebra) Let x be the number of senior and y be the number of junior monks. There are 100 monks in total, so

$$x + y = 100.$$

4 junior monks eat 1 steamed bun, so one junior monk eats

$$1 \div 4 = 0.25$$

buns. Since we also know a senior monk eats 4 buns,

$$4 \times x + 0.25 \times y = 100.$$

Therefore we have the system of equations

$$\begin{cases} x + y & = & 100, \\ 4x + 0.25y & = & 100. \end{cases}$$

Multiplying the second equation by 4 we get

$$16x + y = 400.$$

Subtracting the first equation from this gives us

$$15x = 300,$$

so we solve for x,

$$x = \frac{300}{15} = 20.$$

Plugging back into the first equation,

$$20 + y = 100,$$

so

$$y = 100 - 20 = 80.$$

Thus there are 20 senior monks and 80 junior monks.

Problem 3.50 **A large bottle can hold** 4 **liters of oil, while every two small bottle can hold** 1 **liter of oil. A store has** 100 **liters of oil and the oil exactly fills up** 60 **bottles. How many of each kind of bottle does the store have?**

Answer

20 large, 40 small

Solution 1

Since two small bottles hold 1 liter of oil, each small bottle holds

$$1 \div 2 = 0.5$$

liters of oil. If the store uses only small bottles, it can only hold

$$60 \times 0.5 = 30$$

liters of oil. This is

$$100 - 30 = 70$$

less than they need to hold. Each large bottle holds

$$4 - 0.5 = 3.5$$

extra liters of oil. Hence if they switch

$$70 \div 3.5 = 20$$

bottles from small to large they can hold the correct amount of oil with 60 bottles. hence there are

$$60 - 20 = 40$$

small bottles and 20 large bottles.

Solution 2

(Algebra) Let x be the number of large bottles and y be the number of small bottles. There are 60 bottles in total, so

$$x + y = 60.$$

Since two small bottles hold 1 liter of oil, each holds

$$1 \div 2 = 0.5$$

liters of oil. There are 100 liters of oil in total, so

$$4 \times x + 0.5 \times y = 100.$$

Hence we need to solve the two variable two equation system,

$$\begin{cases} x + y & = & 60, \\ 4x + 0.5y & = & 100. \end{cases}$$

Doubling the second equation we have

$$8x + y = 200,$$

so subtracting the first equation from this gives us

$$7x = 200 - 60 = 140.$$

Hence

$$x\frac{140}{7} = 20,$$

so substituting

$$20 + y = 100$$

and

$$y = 100 - 20 = 80.$$

Hence the store needs 20 large bottles and 80 small bottles.

Problem 3.51 **Charles and David are fast typists. Charles can type 12 more words per minute than David. Charles started typing, and 2 minutes later David also started typing, and they both stopped after 3 more minutes. Given that they typed 780 words altogether, how many words did each of them type?**

Answer

Charles 510, David 270

Solution 1

First note that David types of 3 minutes, and since Charles started 2 minutes before David, Charles types for a total of

$$2 + 3 = 5$$

minutes. Since Charles types 12 words per minute faster than David, he can type

$$5 \times 12 = 60$$

more words than David in 5 minutes. Hence, if instead of Charles typing for 5 minutes and David typing for 3 minutes, David types for

$$5 + 3 = 8$$

minutes, he can type

$$780 - 60 = 720$$

words. Thus, David types

$$720 \div 8 = 90$$

words per minute. Since David actually types for 3 minutes, he types

$$3 \times 90 = 270$$

words in total. Lastly, we have Charles types the remaining

$$780 - 270 = 510$$

words.

Solution 2

(Altebra) Let x be the number of words David can type per minute. Since Charles types 12 more words per minute than David, we know Charles types $x + 12$ words per minute. David types a total of 3 minutes and Charles types a total of

$$2 + 3 = 5$$

minutes to type 780 words, so we have

$$5 \times (x + 12) + 3 \times x = 780.$$

Distributing and combining like terms we have

$$8x = 720$$

so

$$x = \frac{720}{8} = 90$$

and David types 90 words per minute. Thus Charles types

$$90 + 12 = 102$$

words per minute. Hence in total Charles types

$$102 \times 5 = 510$$

words and David types

$$90 \times 3 = 270$$

words.

Problem 3.52 **In the warehouse there were 3 times as much apples as bananas at the beginning. Suppose 250 pounds of bananas and 600 pounds of apples were sold every day, and a few days later the bananas were sold out and 750 pounds of apples were left. How many pounds of each fruit were there originally?**

Answer

1250 pounds of bananas, 3750 pounds of apples

Solution 1

We know that the original ratio of apples to bananas is 3 : 1. 600 pounds of apples are sold every day. Since

$$3 : 1 = 3 \times 200 : 1 \times 200 = 600 : 200$$

if we assume that only 200 pounds of bananas were sold each day, then the amounts of apples and bananas in the warehouse would always be kept at the ratio 3 : 1. Then at the end, there were 750 pounds of apples, so there would have been

$$750 \div 3 = 250$$

pounds of bananas remaining. Since each day

$$250 - 200 = 50$$

pounds less of bananas were sold than the actual amount, the number of days was

$$250 \div 50 = 5.$$

Thus at the beginning there were

$$250 \times 5 = 1250$$

pounds of bananas, and

$$1250 \times 3 = 3750$$

pounds of apples.

Solution 2

(Algebra) Let x be the number pounds of bananas at the beginning. Since there are 3 times as many apples, there are $3x$ pounds of apples. Let y be the number of days it takes to sell all the bananas. Since 250 pounds of bananas are sold every day, we have

$$x - 250 \times y = 0.$$

600 pounds of apples are sold every day, but 750 pounds remain after y days, so

$$3x - 600 \times y = 750.$$

This gives the system of equations

$$\begin{cases} x - 250y &= 0, \\ 3x - 600y &= 750. \end{cases}$$

Multiplying the first equation by three we have

$$3x - 750y = 0$$

and subtracting this from the second equation gives us

$$150y = 750$$

and

$$y = \frac{750}{150} = 5.$$

Substituting back into the first equation,

$$x - 250 \times 5 = 0$$

so

$$x = 1250.$$

Hence there are 1250 pounds of bananas and

$$3 \times 1250 = 3750$$

pounds of apples at the beginning.

Problem 3.53 **Jack went hiking. His uphill speed was 3 miles per hour and downhill speed was 5 miles per hour, and he hiked a total of 6 hours including both uphill and downhill, with total distance 23 miles. How many hours did he spent uphill and downhill respectively?**

Answer

3.5 hours uphill, 2.5 hours downhill

Solution 1

If we suppose Jack walk for the whole 6 hours at a speed of 3 miles per hour, he would travel a total of

$$3 \times 6 = 18$$

miles, which is

$$23 - 18 = 5$$

miles less than he actually hiked. Every hour Jack spends hiking downhill at 5 miles per hour allows him to travel an additional

$$5 - 3 = 2$$

miles than he would hiking uphill. Therefore, Jack must spend

$$5 \div 2 = 2.5$$

hours hiking downhill. Hence he spends the other

$$6 - 2.5 = 3.5$$

hours hiking uphill.

Solution 2

(Algebra) Let x be the number of hours Jack hikes uphill and y the number downhill. He hikes a total of 6 hours, so

$$x + y = 6.$$

We also know Jack's hike was 23 miles long, so as he travels 3 miles per hour uphill and 5 miles per hour downhill

$$3 \times x + 5 \times y = 23.$$

We therefore want to solve the system of equations

$$\begin{cases} x+y &= 6, \\ 3x+5y &= 23. \end{cases}$$

Multiplying the first equation by 3 we have

$$3x+3y = 18.$$

Subtracting this from the second equation,

$$2y = 5$$

so

$$y = \frac{5}{2} = 2.5.$$

Substituting back into the first equation

$$x + 2.5 = 6$$

we get

$$x = 6 - 2.5 = 3.5.$$

Hence Jack spends 3.5 hours hiking uphill and 2.5 hours hiking downhill.

Problem 3.54 **Lily spent $490 to buy 80 color pencils for her art class, including red, green. and blue colors. The red pencils cost $2 each, the green ones cost $5 each, and the blue ones cost $10 each. Suppose she bought the same number of green and blue pencils. How many of each type of pencils did she buy?**

Answer

20 red, 30 green, 30 blue

Solution 1

Since Lily bought the same number of green and blue pencils, we can pretend that one green pencil and one blue pencil come bundled as a green-blue pair that costs

$$5 + 10 = 15$$

dollars. If we suppose all the pencils Lily bought were red, then the total cost would have been

$$80 \times 2 = 160$$

dollars. The actual amount 490 is

$$490 - 160 = 330$$

dollars more. Replacing a pair of red pencils, which costs

$$2 \times 2 = 4$$

dollars, with a green-blue pair costs, which costs 15 dollars, is

$$15 - 4 = 11$$

dollars more expensive. Therefore the total number of green-blue pairs is

$$330 \div 11 = 30.$$

Thus there were 30 green and 30 blue pencils. The remaining

$$80 - 30 - 30 = 20$$

are red pencils.

Solution 2

(Algebra) Let x be the number of red pencils and y be the number of green pencils that Lily bought. Thus Lily also buys y blue pencils. Since she buys 80 in total, we have

$$x + y + y = 80.$$

Each red pencil costs \$2, each green pencil \$5, and each blue pencil \$10. Since Lily spends \$490 in total, we have

$$x \times 2 + y \times 5 + y \times 10 = 490.$$

Combining like terms in the two equations above gives us the system of equations

$$\begin{cases} x + 2y = 80, \\ 2x + 15y = 490. \end{cases}$$

Multiplying the first equation by 2 we have

$$2x + 4y = 160,$$

and subtracting this from the second equation gives us

$$11y = 330.$$

Therefore

$$y = \frac{330}{11} = 30.$$

Substituting back into the first equation, we know

$$x + 2 \times 30 = 80$$

so we can solve for x to get

$$x = 80 - 60 = 20.$$

Hence Lily bought 20 red, 30 green, and 30 blue pencils.

Problem 3.55   **A spider has 8 legs and no wings. A dragonfly has 6 legs and 2 pairs of wings. A cicada has 6 legs and one pair of wings. There are a total of 18 bugs of these types, with 118 legs and 20 pairs of wings. How many dragonflies are there?**

Answer

7

Solution 1

There are 18 bugs in total. Spiders have no wings, so let's first focus on the number of legs. If all the bugs are spiders, there are a total of

$$18 \times 8 = 144$$

legs, which is

$$144 - 118 = 26$$

more than the actual amount. Each spider has

$$8 - 6 = 2$$

more legs than a dragonfly or cicada, so there must be a combined total of

$$26 \div 2 = 13$$

dragonflies and cicadas. Each cicada has 1 pair of wings and each dragonfly has 2 pairs of wings, so since we need 20 wings in total, we must have

$$20 - 13 = 7$$

extra pairs of wings, so a total of 7 dragonflies.

Solution 2

(Algebra) Let x be the number of spiders, y the number of dragonflies, and z the number of cicadas. There are 18 total bugs, so

$$x + y + z = 18.$$

We also know there are 118 total legs, so counting up the legs of each type we must have

$$8 \times x + 6 \times y + 6 \times z = 118.$$

Finally, for the number of pairs of wings we have

$$0 \times x + 2 \times y + 1 \times z = 20.$$

Hence we need to solve the system of equations

$$\begin{cases} x + y + z & = & 18, \\ 8x + 6y + 6z & = & 118, \\ 2y + z & = & 20. \end{cases}$$

Multiplying the first equation by 8 we have

$$8x + 8y + 8z = 144.$$

Subtracting the second equation from this gives us that

$$2y + 2z = 26$$

and dividing this by 2 we have

$$y + z = 13.$$

Subtracting this from the third equation in our system gives us

$$y = 20 - 13 = 7,$$

the number of dragonflies.

Problem 3.56 **Some friends rent some boats. If** 4 **people get in each boat,** 8 **people will not fit in the boats. If** 5 **people get in each boat,** 6 **people will not fit in the boats. How many friends are there? How many boats?**

Answer

16 friends, 2 boats

Solution 1

Note for each 4 person boat switched to a 5 person boat,

$$5 - 4 = 1$$

extra person can get on the boat. Since switching from 4 person boats to 5 person boats allows

$$8 - 6 = 2$$

extra people to fit in the boats, there must be 2 boats. If 5 people get in each boat, the boats hold

$$2 \times 5 = 10$$

people with 6 left over. Hence there are

$$10 + 6 = 16$$

people in total.

Solution 2

(Algebra) Let x be the number of boats. In the first scenario 4 people fit in each boat with 8 left over, so the total number of people is

$$4 \times x + 8.$$

We also know that if the same people fit 5 in a boat there are 6 left over, so the number of people is also equal to

$$5 \times x + 6.$$

Thus,
$$4x + 8 = 5x + 6$$

and combining like terms,

$$x = 8 - 6 = 2$$

so there are 2 boats. Hence there are

$$4 \times 2 + 8 = 5 \times 2 + 6 = 16$$

people in total.

Problem 3.57 **A grocer mixed grape juice, which costs \$2.25 per gallon, with cranberry juice, which costs \$1.75 per gallon. How many gallons of each should be used to make 200 gallons of a cranberry/grape juice mix that costs \$2.10 per gallon?**

Answer

Grape: 140, Cranberry: 60

Solution 1

The grocer wants a mix of 200 gallons that costs \$2.10 per gallon. This is a total cost of
$$200 \times 2.10 = 420$$

dollars. If the grocer uses 200 gallons of cranberry juice, it will cost

$$200 \times 1.75 = 350$$

dollars which is

$$420 - 350 = 70$$

dollars cheaper than the needed price. Switching one gallon from cranberry to grape juice costs an extra

$$2.25 - 1.75 = 0.5$$

dollars. Hence if they switch

$$70 \div 0.5 = 140$$

gallons to grape juice the grocer will have the correct mixture. Hence they need 140 gallons of grape juice and

$$200 - 140 = 60$$

gallons of cranberry juice.

Solution 2

(Algebra) Assume there are x gallons of grape juice and y gallons of cranberry juice. There are 200 gallons in total so

$$x + y = 200.$$

The target mixture costs \$2.10 per gallon, so 200 gallons should cost

$$200 \times 2.10 = 420$$

dollars. Therefore

$$2.25 \times x + 1.75 \times y = 420$$

as grape juice costs \$2.25 per gallon and cranberry juice \$1.75 per gallon. We are left with the system of equations

$$\begin{cases} x + y & = & 200, \\ 2.25x + 1.75y & = & 420. \end{cases}$$

Solving the first equation for y,

$$y = 200 - x$$

so substituting this into the second equation we have

$$2.25x + 1.75 \times (200 - x) = 420$$

or

$$2.25x + 350 - 1.75x = 420.$$

Combining like terms gives us

$$0.5x = 70$$

so doubling both sides,

$$x = 140.$$

Lastly, we substitute back into the first equation so

$$y = 200 - 140 = 60.$$

Hence the grocer should use 140 gallons of grape juice and 60 gallons of cranberry juice.

Problem 3.58 **The owner of the Fancy Food Shoppe wishes to mix cashews selling at \$8.00 per kilogram and pecans selling at \$7.00 per kilogram. How much of each kind of nut should be mixed to get 8 kg worth \$7.25 per kilogram?**

Answer

Cashews: 2 kg, Pecans: 6 kg

Solution 1

Since the owner wants the 8 kg mixture to be worth \$7.25 per kilogram, in total it should cost

$$8 \times 7.25 = 58$$

dollars. If the owner gets all 8 kilograms from pecans, the mixture would cost

$$8 \times 7.00 = 56$$

dollars,

$$58 - 56 = 2$$

less than the actual cost. The extra \$2 comes from cashews. Each kilogram of cashews costs

$$8.00 - 7.00 = 1$$

dollar more than a kilogram of pecans, and replacing 2 pounds of pecans with cashews will increase the price of the mixture to the correct \$58. The remaining

$$8 - 2 = 6$$

kilograms are from pecans.

Solution 2

(Algebra) Assume there are x kilograms of cashews and y kilograms of pecans. We first have

$$x + y = 8$$

as the owner want 8 kg in total. The 8 kg mixture should costs \$7.25 per kg, or

$$8 \times 7.25 = 58$$

dollars in total. Hence

$$8 \times x + 7 \times y = 58$$

and we have the system of equations

$$\begin{cases} x + y & = & 8, \\ 8x + 7y & = & 58. \end{cases}$$

Multiply the first equation by 7 to get

$$7x + 7x = 56$$

so if we subtract this from the second equation we have

$$x = 58 - 56 = 2.$$

Substituting back into the first equation,

$$2 + y = 8$$

so

$$y = 8 - 2 = 6.$$

Therefore the owner should use 2 kilograms of cashews and 6 kilograms of pecans.

Problem 3.59 A convenience store owner wishes to mix together raisins and roasted peanuts to produce a high energy snack for hikers. The raisins sell for $3.50 per kilogram and the nuts sell for $4.75 per kilogram. How many kilograms of each should be mixed together to obtain 20 kg of this snack with a price of $4.00 per kilogram?

Answer

Raisons: 12, Peanuts: 8

Solution 1

20 kg of a snack that is $4.00 per kg should cost

$$20 \times 4.00 = 80$$

dollars. If the store owner uses 20 kilograms from only raisins, the 20 kg would only cost,

$$20 \times 3.50 = 70$$

dollars. This is

$$80 - 70 = 10$$

cheaper than it should be. Each kilogram of peanuts costs

$$4.75 - 3.50 = 1.25$$

more than a kilogram of raisins. Thus, there must be

$$10 \div 1.25 = 8$$

kilograms of peanuts. The remaining

$$20 - 8 = 12$$

kilograms are raisins.

Solution 2

(Algebra) Assume there are x kilograms of raisins and y kilograms of peanuts, so

$$x + y = 20$$

if the store owner wants to create a 20 kg mixture. This mixture should cost $4.00 per kg so 20 kg will cost

$$20 \times 4.00 = 80$$

dollars. Since this cost is a combination of raisins and peanuts,

$$3.50 \times x + 4.75 \times y = 80.$$

These give us the system of equations

$$\begin{cases} x + y & = 20, \\ 3.5x + 4.75y & = 80. \end{cases}$$

Solving the first equation for x,

$$x = 20 - y$$

so substituting this into the second equation we have

$$3.5 \times (20 - y) + 4.75y = 80.$$

Distributing and combining like terms we have

$$1.25y = 10$$

so

$$y = \frac{10}{1.25} = 8.$$

Plugging this value in for y gives

$$x = 20 - 8 = 12.$$

Therefore the convenience store owner should use 12 kilograms of raisins and 8 kilograms of peanuts.

Problem 3.60 **A meat distributor paid $2.50 per pound for hamburger meat and $4.50 per pound for ground sirloin. How many pounds of each did he use to make 100 pounds of meat mixture that will cost $3.24 per pound?**

Answer

37 pounds sirloin, 63 pounds hamburger

Solution 1

The meat distributor wants a mixture that costs $3.24 per pound, so 100 pounds will cost

$$100 \times 3.24 = 324$$

dollars. 100 pounds of ground sirloin costs

$$100 \times 4.50 = 450$$

dollars, which is

$$450 - 324 = 126$$

dollars more expensive than the goal. Each pound of hamburger meat is

$$4.50 - 2.50 = 2$$

dollars cheaper than a pound of ground sirloin, so replacing

$$126 \div 2 = 63$$

pounds of ground sirloin with hamburger meat will reduce the price to $324 as needed. Hence the distributor used

$$100 - 63 = 37$$

pounds of ground sirloin and 63 pounds of hamburger meat.

Solution 2

(Algebra) Assume there are x pounds of hamburger meat and y pounds of ground sirloin. The total mixture is 100 pounds so

$$x + y = 100.$$

The meat distributor wants an entire 100 pound mixture that costs $3.24 per pound, which costs

$$100 \times 3.24 = 324$$

dollars in total. Since this mixture is a combination of hamburger meat and ground sirloin,

$$2.50 \times x + 4.50 \times y = 324.$$

This gives the system of equations

$$\begin{cases} x+y &= 100, \\ 2.5x+4.5y &= 324. \end{cases}$$

Solving the first equation for y,

$$y = 100 - x.$$

Substituting into the second equation,

$$2.5x + 4.5 \times (100 - x) = 324$$

so distributing we have

$$2.5x + 450 - 4.5x = 324.$$

Combining like terms we get

$$-2x = -126$$

so

$$x = \frac{-126}{-2} = 63.$$

Lastly we solve for y,

$$y = 100 - 63 = 37.$$

Hence the meat distributor should use 37 pounds of hamburger meat and 63 pounds of ground sirloin.

4. Motion—Speed, Time, and Distance

Problem 4.1 **Suppose a train travels a distance of 120 miles in 3 hours. What is the average speed of the train?**

Answer

40 miles per hour

Solution

The average speed is total distance divided by total time, or

$$120 \div 3 = 40 \text{ miles per hour (mph)}.$$

Problem 4.2 **Mary and Amy rollerblade at an average speed of 9 miles per hour for 3.5 hours, how far will they travel?**

Answer

31.5 miles

Solution

We know the speed is 9 miles/hr and the total time they travel is 3.5 hours, so the total distance traveled is

$$9 \times 3.5 = 31.5 \text{ miles.}$$

Problem 4.3 In a cross-country race, Tony drove his car for 707 kilometers (km) in 7 hours. What was his average speed?

Answer

101 km/h

Solution

The total distance is 707 km and the total time he used is 7 hours, so his average speed is

$$707 \div 7 = 101 \text{ km/h.}$$

Problem 4.4 A tennis ball is thrown a distance of 20 meters. What is its speed if it takes 0.5 seconds to cover the distance?

Answer

40 m/s

Solution

To find the average speed, we need to use the total distance divided by the total time, then

$$20 \div 0.5 = 40$$

gives us the average speed in m/s.

Problem 4.5 A bat is flying at a speed of 45 kilometers per hour. How much time does it take to travel a distance of 1,800 kilometers?

Answer

40 hours

Solution

To find the amount of time it takes, we use the total distance divided by the speed, which is,

$$1800 \div 45 = 40 \text{ hours.}$$

Problem 4.6 **Adam takes a train to go visit a friend who lives in a city that is 360 kilometers away. The train left his home station at 8:35AM, and arrived at the destination station at 1:05PM. How fast has the train traveled measured by average speed?**

Answer

80 km/h

Solution

The total time Adam traveled is 4 hours and 30 minutes, which is 4.5 hours.

Since the train traveled a total of 360 km, the train has an average speed

$$360 \div 4.5 = 80 \text{ km per hour (km/h).}$$

Problem 4.7 **Thomas and his family went on a road trip last week. They traveled 50 mph from Chicago, IL to Minneapolis, MN and 65 mph on the return trip. What was the average speed for the entire round trip?**

Answer

$\dfrac{1300}{23}$ miles per hour

Solution

Since we are not given the distance in the problem, we can assume the distance is 650 miles. So the time takes Thomas and his family to get to Minneapolis, MN is

$$650 \div 50 = 13 \text{ hours,}$$

and the return trip takes

$$650 \div 65 = 10 \text{ hours.}$$

For the round trip, the total distance is 1300 miles and the total time spent is

$$13 + 10 = 23 \text{ hours.}$$

To find the average speed, we calculate the total distance divided by the total time:

$$1300 \div 23 = \frac{1300}{23} \text{ miles per hour,}$$

which is approximately 56.5 miles per hour.

Problem 4.8 **Jimmy and his brother took a circular ride at an amusement park on averages of 30 miles per hour and took them $2\frac{1}{2}$ minutes. Roughly how big is the diameter of the circular track?**

Answer

1.25 miles

Solution

In order to find the distance they traveled, all we need to do is to multiply the speed and the amount of time they used. However, the unit of the speed is miles per hour, while the unit of the time is minutes. Now, we need to convert one of the unit, so that we can multiple them.

$$2.5 \text{ minutes } = 2.5 \times \frac{1}{60} = \frac{1}{24} \text{ hours.}$$

Now we can find the distance,

$$30 \times \frac{1}{24} = 1.25 \text{ miles.}$$

We are asked to find the rough measurement of the diameter of the track. Let's assume the track is a perfect circle. The formula for perimeter of the circle is the diameter multiplied by a constant π. So the diameter $= \frac{1.25}{\pi} \approx 0.398$ miles, which is approximately 2100 feet.

Problem 4.9 **If I drive from Irvine to Fullerton at 60 miles per hour and then from Fullerton to Irvine at 40 miles per hour, what is my average speed for the whole journey?**

Answer

48 miles per hour

Solution

Since the distance is the same in both directions, we can assume that the distance is a value that is easy to deal with. Therefore, assume the distance between Irvine and Fullerton is 120 miles. (I know it may seem ridiculous because it's not that far from Irvine to Fullerton, but a good question to think about is whether it matters if you assume the distance 120 miles or 20 miles.) So the time takes me to reach Fullerton is

$$120 \div 60 = 1/3 \text{ hours,}$$

and the return trip takes

$$120 \div 40 = 3 \text{ hours.}$$

For the round trip, the total distance is 240 miles and the total time spent is

$$2 + 3 = 5 \text{ hours.}$$

To find the average speed, we can calculate the total distance divided by the total time

$$240 \div 5 = 48$$

miles per hour.

Problem 4.10 **Katie went hiking on a hill near her home. From the bottom of the hill, She went up to the top and then came down along the same trail, back to the spot she started. Assume her uphill speed was 3 miles per hour, and her downhill speed was 6 miles per hour. What is her average speed for the whole uphill-downhill trip?**

Answer

4 miles per hour.

Solution

We don't know the length of the trail, or the time she spent uphill or downhill. However, we may assume that she took 2 hours going uphill. Then the trail length was $3 \times 3 = 6$ miles, and the downhill journey took $6 \div 6 = 1$ hour. Therefore the total time was $2 + 1 = 3$ hours, and the average speed of the whole trip was $(6 + 6) \div 3 = 4$ miles per hour.

Note: what if we assume a different number of hours for her uphill trip? Will the answer be the same? Why?

Problem 4.11 **Melisa drove for 3 hours at a rate of 50 miles per hour and for 2 hours at 60 miles per hour. What was her average speed for the whole journey?**

Answer

54 miles per hour

Solution

First we need to find the total distance Melisa traveled, which is

$$3 \times 50 + 2 \times 60 = 270 \text{ miles.}$$

We also need to find the total amount of time she traveled,

$$2 + 3 = 5 \text{ hours.}$$

We can then find Melisa's average speed, which is

$$270 \div 5 = 54 \text{ miles per hour.}$$

Problem 4.12 **Frank drives his car for a distance of** 300 **miles. For the first** 135 **miles, he drives at a constant speed of** 45 **miles per hour. At what constant speed does he drive for the remaining distance to average** 50 **miles per hour?**

Answer

55 miles per hour.

Solution

For the total distance 300 miles, Frank has an average speed of 50 miles per hour, therefore the total time he takes driving is $\dfrac{300}{50} = 6$ hours. For the first 135 miles, he drives at 45 miles per hour, thus the time he takes for that part is $\dfrac{135}{45} = 3$ hours. After the first 135 miles, there are $300 - 135 = 165$ miles, and it takes him the remaining 3 hours to cover it. That means the remaining distance requires $\dfrac{165}{3} = 55$ miles per hour.

Problem 4.13 **Suppose a truck travels in segments that are described in the table below:**

Segment	Distance (miles)	Time (hours)
1	30	1
2	90	2
3	50	1

What is the average speed of the truck?

Answer

42.5 miles/hr

Solution

First, we need to find the total distance the truck travels, which is

$$30 + 90 + 50 = 170 \text{ miles.}$$

Then the total hours the truck traveled is

$$1 + 2 + 1 = 4 \text{ hours.}$$

hours. So, the average speed of the truck is

$$170 \div 4 = 42.5 \text{ miles per hour.}$$

Problem 4.14 **Joe and his family are planning to go to a national park which is 600 miles away from the home. How fast in miles per hour must they drive if they want to get there in 15 hours?**

Answer

40 miles per hour

Solution

To find the speed, use the total distance divides by the total time, which is

$$600 \div 15 = 40 \text{ miles per hour.}$$

Problem 4.15 **Ling goes mountain hiking in a park. She first walks uphill at a speed of 2.5 miles per hour, and she next walks downhill at a speed of 4 miles per hour. The round trip takes 3.9 hours. What is the distance for the round trip?**

Answer

12 miles

Solution 1

Since Ling walks the same distance up and down the mountain, let's assume the distance is 10 miles. So the time it take Ling to reach the top of the mountain is

$$10 \div 2.5 = 4 \text{ hours},$$

and the walk back downhill takes

$$10 \div 4 = 2.5 \text{ hours}.$$

The round trip therefore takes

$$4 + 2.5 = 6.5 \text{ hours}.$$

To find the average speed, we calculate the total distance divided by the total time. The total distance is 20 miles and the total time is 6.5 hours, so the average speed is $20 \div 6.5 = \dfrac{20}{6.5}$ miles per hour. Since the round trip takes 3.9 hours, we can find the distance $\dfrac{20}{6.5} \times 3.9 = 12$ miles for the round trip.

Solution 2

Let x be the distance from the bottom to the top of the mountain. The time it take Ling to reach the top of the mountain is

$$x \div 2.5 = \frac{2x}{5} \text{ hours,}$$

and the walk back downhill takes

$$x \div 4 = \frac{x}{4} \text{ hours.}$$

The round trip therefore takes

$$\frac{2x}{5} + \frac{x}{4} = \frac{13x}{20} \text{ hours.}$$

Therefore

$$\frac{13x}{20} = 3.9$$

so solving for x we get

$$x = 6.$$

Since this is the distance one way, the round trip is 12 miles.

Problem 4.16 **Stephanie begins walking at a pace of 4 km per hour from one end of the trail that is 34 km long. Bob begins at the other end of the trail at the same time, walking towards Stephanie at a pace of 6 km. How long will it take for them to pass each other?**

Answer

3.4 hours

Solution 1

(Relative Speed) Since Stephanie and Bob are traveling in opposite directions, their relative speed is the sum of the two speeds, which is,

$$4 + 6 = 10 \text{ km per hour}$$

towards each other. We can then find the time, which is

$$34 \div 10 = 3.4 \text{ hours}$$

until they pass each other.

Solution 2

(Algebra) Let the time it takes for them to pass each other to be x hours. In x hours, Stephanie travels $4 \times x$ miles and Bob travels $6 \times x$ miles. If they pass each other after exactly x hours,

$$4x + 6x = 34.$$

Solve for x, then

$$x = 3.4,$$

so it takes 3.4 hours for them to pass each other.

Problem 4.17 **Terry and Susan are entered in a 24-mile race. Susan's average rate is 4 miles per hour and Terry's average rate is 6 miles per hour. Both start at the same time. How far will Susan be away from the finish line when Terry crosses the line?**

Answer

8 miles

Solution

First find the time that takes Terry to finish the race, which is

$$24 \div 6 = 4 \text{ hours.}$$

After 4 hours of running, the distance that Susan has run is

$$4 \times 4 = 16 \text{ miles.}$$

Since the race is 24 miles long, Susan still has

$$24 - 16 = 8 \text{ miles}$$

until she reaches the finish line.

Problem 4.18 **John's house and Mary's house are 14 miles apart. They start at noon to walk toward each other in order to go to a book fair together. John walks at a rate of 3 mph, and Mary walks at a rate of 4 mph. How many hours will it take them to meet?**

Answer

2 hours

Solution 1

(Use Relative Speed) John and Mary are traveling in opposite directions toward each other, therefore, their relative speed is the sum of their two speeds, which is $3 + 4 = 7$ miles per hour towards each other. Therefore, it takes them $14 \div 7 = 2$ hours until they meet.

Solution 2

(Algebra) Let the time it takes for them to meet each other be x hours. In x hours, John travels

$$3 \times x$$

and Mary travels

$$4 \times x$$

miles, so if they meet up at that time,

$$3x + 4x = 14.$$

Solve for x, we have

$$x = 2,$$

so it takes 2 hours for them to meet each other.

Problem 4.19 **Jasmine took a walk after dinner. She first walked 5 km in 1.5 hours, and then walked for 1 km in 0.5 hour in the same direction. What is her average speed for the whole journey?**

Answer

3 km/h

Solution

In order to find the average speed, we need to first find the total distance, which is

$$5 + 1 = 6 \text{ km},$$

and the total time, which is

$$1.5 + 0.5 = 2 \text{ hours}.$$

Then, Jasmine's average speed

$$6 \div 2 = 3 \text{ km/h}.$$

Problem 4.20 **Two friends leave the same place at the same time traveling in the same direction. One travels at a speed of 55 mph and the other travels at a rate of 65 mph. After 2 hours, how many miles will they be away from each other?**

Answer

20 miles

Solution 1

(Use Relative Speed) Since the two friends are traveling in same direction, their relative speed is the difference of the two speeds, which is

$$65 - 55 = 10 \text{ miles per hour}$$

away from each other. After 2 hours of traveling, the difference in distance becomes,

$$10 \times 2 = 20 \text{ miles}.$$

Hence, they are 20 miles away from each other after 2 hours.

Solution 2

We can calculate the distances for both of the two friends after 2 hours, and then subtract them to find the difference in distance. The first friend travels

$$65 \times 2 = 130 \text{ miles}$$

in the two hours while the second travels

$$55 \times 2 = 110 \text{ miles.}$$

miles. The difference,

$$130 - 110 = 20 \text{ miles,}$$

is how far apart they are after 2 hours.

Problem 4.21 **Mike drives his car for a round trip between LA and San Diego. He drives at 70 miles per hour to get from LA to San Diego. At what speed should he drive back, if his average speed for the round trip is 60 miles per hour?**

Answer

$\dfrac{105}{2}$ miles per hour

Solution

We are not given any distances, so assume the distance between LA and San Diego is 210 miles, so the round trip distance is 420 miles. If the average speed for the round trip is 60 miles per hour, the round trip will take

$$420 \div 60 = 7 \text{ hours.}$$

hours. Then, we can find the time it took Mike to travel from LA to San Diego, which is

$$210 \div 70 = 3 \text{ hours.}$$

So the time he used to drive back is

$$7 - 3 = 4 \text{ hours.}$$

Therefore, the speed he must drive back is

$$210 \div 4 = \frac{105}{2} \text{ miles per hours.}$$

Problem 4.22 **Alice leaves site A toward site B at the same time Bob leaves site B toward A. Alice drives at 40 miles per hour, and Bob drives at 60 miles per hour. After they pass by each other, Alice drives 4.5 additional hours to arrive at B. How far is it between A and B?**

Answer

300 miles

Solution

Alice drives an additional 4.5 hours after they meet, so in that time she travels

$$40 \times 4.5 = 180 \text{ miles}$$

to get to site B. It takes Bob

$$180 \div 60 = 3 \text{ hours}$$

to travel these same 180 miles. Since Alice and Bob left at the same time, we know Alice also drove 3 hours from A to the place they met, which makes that distance

$$40 \times 3 = 120 \text{ miles.}$$

So the total distance from A to B is

$$120 + 180 = 300 \text{ miles.}$$

Problem 4.23 **Linda goes mountain biking in a park. She first bikes on flat road at a speed of 12 miles per hour, then goes uphill at a speed of 9 miles per hour, and she next bikes downhill at a speed of 18 miles per hour, along the same trail as the uphill trip. Finally she goes back home along the same flat road she traveled earlier. The round trip takes 4 hours. What's the distance for the round trip?**

Answer

48 miles

Solution

Although we don't know the distance of each segment of the trip, we do know that the uphill and downhill distances are the same, so we can calculate the average speed of the uphill and downhill segments. For easier calculation, assume the uphill distance is 18 miles (it would give us the same average speed if we chose any other value), so the downhill distance is also 18 miles. Thus the uphill time is $18 \div 9 = 2$ hours and the downhill time is $18 \div 18 = 1$ hour. So the average speed of the uphill and downhill portion is $(18 + 18) \div (2 + 1) = 12$ miles per hour. This coincides with the flat road speed. Therefore, the average speed of the whole trip is also 12 miles per hour. Since it takes 4 hours to complete the round trip, the total distance is $12 \times 4 = 48$ miles.

Note: in this problem, the average speed of the uphill and downhill trip coincides with the flat road speed. If they didn't coincide, this solution would not work, and there would not be enough information for a definite answer.

Problem 4.24   As Emily is riding her bicycle on a long straight road, she spots Emerson skating in the same direction 1/2 mile in front of her. After she passes him, she can see him in her rear view mirror until he is 1/2 mile behind her. Emily rides at a constant rate of 12 miles per hour, and Emerson skates at a constant rate of 8 miles per hour. For how many minutes can Emily see Emerson?

Answer

15

Solution

Since they are traveling in the same direction, their relative speed is the difference of the two speeds,

$$12 - 8 = 4 \text{ miles per hour}$$

towards each other before passing, and away from each other after passing. Emerson starts a half mile in front of Emily, and ends a half mile behind Emily, so Emily travels a total of 1 mile in reference to Emily. Therefore, the time Emerson is in her view is

$$1 \div 4 = \frac{1}{4}$$

of an hour, which is 15 minutes.

Problem 4.25   Two trains, 121 meters and 99 meters in length respectively, are moving in opposite directions, one at the rate of 40 km/h and the other at the rate of 32 km/h. In what time will they be completely clear of each other from the moment they meet?

Answer

11 seconds

Solution

Since the two trains are moving in opposite directions, then the relative speed is the sum of the two speeds,

$$40 + 32 = 72 \text{ km/h}$$

In order for the two trains to be completely clear of each other from the moment they meet, the total distance the two trains traveled is the sum of the length of the two trains, which is

$$121 + 99 = 220 \text{ meters.}$$

Notice that the units for the speed and the length of the train are not consistent. Hence, we need to convert one to the other before we go forward. One way is to change the speed from km/h to m/s. We know that 1 km = 1000 m and 1 hour = 3600 seconds, so 72 km/h is

$$72 \times 1000 \div 3600 = 20 \text{ m/s.}$$

Now we can find the time it takes the trains to complete clear each other:

$$220 \div 20 = 11 \text{ seconds.}$$

Problem 4.26 **Two trains, 121 meters and 99 meters in length respectively, are moving in the same direction, one at the rate of 40 km/h and the other at the rate of 32 km/h. How long would it take to be completely clear of one another, if the faster train has just met up with the back of the slower train.**

Answer

99 seconds

Solution

The two trains are moving in the same direction, so the relative speed is the difference of the two speeds, which is

$$40 - 32 = 8$$

km/h. Changing this speed into m/s (using 1 km = 1000 m and 1 hr = 3600 s),

$$8 \times 1000 \div 3600 = \frac{20}{9}$$

m/s. In order to completely clear each other, the total distance is the sum of the length of the two trains, which is

$$121 + 99 = 220$$

meters. We can then find the time it takes to do so, which is

$$220 \div \frac{20}{9} = 99$$

seconds.

Problem 4.27   **A train, 110 meters in length, travels at 60km/h. In what time will it pass a man who is walking at 6km/h (i) against it; (ii) in the same direction?**

Answer

(i) 6 seconds, (ii) $\dfrac{22}{3}$ seconds

Solution

Relative to the train, the man is small, so we consider the length of the man to be 0.

(i) Since they are moving in the opposite direction, the relative speed is the sum of the two speeds, which is

$$60 + 6 = 66$$

km/h. Since the length of the train is in meters, we change km/h to m/s (using 1 km = 1000 m and 1 hr = 3600 s), so

$$66 \times 1000 \div 3600 = \frac{55}{3}$$

m/s is the speed of the train. The total distance in this problem is the length of the train, which is 110 meters. Then we can calculate the time,

$$110 \div \frac{55}{3} = 6$$

seconds. Therefore, the train will pass the man in 6 seconds.

(ii) If we instead have the train and man moving in the same direction, the relative speed is the difference of the two speeds, which is

$$60 - 6 = 54$$

km/h. Changing the units as in part (i),

$$54 \times 1000 \div 3600 = 15$$

is the speed of the train in m/s. The total distance is still 110 meters, so it takes the train

$$110 \div 15 = \frac{22}{3} \approx 7.33$$

seconds to pass the man.

Problem 4.28 **A train moving 25 km/h takes 18 seconds to pass a platform. Next, it takes 12 seconds to pass a man walking at 5 km/h in the opposite direction. Find the length of the train and of the platform.**

Answer

Train: 100 meters, Platform: 125 meters

Solution

Notice that the unit for the speed and the unit for the time is not consistent, so we need to convert them in order to move forward. Using 1 km = 1000 m and 1 hr = 3600 s, we can convert km/h into m/s. We have that the train's speed is

$$25 \times 1000 \div 3600 = \frac{125}{18}$$

m/s and the man's speed is

$$5 \times 1000 \div 3600 = \frac{25}{18}$$

m/s. Therefore, the length of the platform is

$$\frac{125}{18} \times 18 = 125$$

meters. Since the train and the man are moving in the opposite direction, their relative speed is the

$$\frac{125}{18} + \frac{25}{18} = \frac{25}{3},$$

measured in m/s. Thus the length of the train is

$$\frac{25}{3} \times 12 = 100$$

meters.

Problem 4.29 It takes 6 hours for an airplane to fly a round trip. If the speed of the airplane is 1500 km per hour on the departure trip, and 1200 km per hour on the return trip. What is the one-way distance ?

Answer

4000 km

Solution 1

(Ratios) The ratio of the speeds of the airplane on the departure and return trips is

$$1500 : 1200 = 5 : 4.$$

Since the distance is the same both ways, the time spend on the trips will be in the opposite ratio $4 : 5$. Since the entire trip is 6 hours,

$$\frac{4}{9} \times 6 = \frac{8}{3}$$

hours will be spend on the departure trip. Therefore, the distance one-way is

$$1500 \times \frac{8}{3} = 4000$$

km.

Solution 2

(Algebra) Let the one-way distance to be x km. Therefore, the departure trip takes

$$x \div 1500 = \frac{x}{1500}$$

hours, while the return trip takes

$$x \div 1200 = \frac{x}{1200}$$

hours. Since the round trip takes 6 hours, we have

$$\frac{x}{1500} + \frac{x}{1200} = 6.$$

We can then solve for x to get

$$x = 4000.$$

Therefore, the distance one-way is 4000 km.

Problem 4.30 **Starting at the same time, Tom and Jerry walk toward each other. Tom walks from site A to B at 5 miles per hour. Jerry walks from B to A. After they meet, Jerry walks an additional 10 miles to arrive at A, and Tom spends additional 1.6 hours to walk and arrive at B. What is the rate at which Jerry walks?**

Answer

4 miles per hour

Solution

After they meet, Jerry walks for an additional 10 miles to get to A. This means Tom walked 10 miles before they met. Since Tom walks at 5 miles per hour, this took him

$$10 \div 5 = 2$$

hours. Since Tom walks an additional 1.6 hours after them meet, he walks

$$1.6 \times 5 = 8$$

additional miles to get to B. Jerry walked these same 8 miles in the 2 hours before they met, so Jerry walks at a speed of

$$8 \div = 2 = 4$$

miles per hour.

Problem 4.31 **Starting at the same time, Heather and Brenda drive their cars from site A toward site B. Heather drives at 52 km per hour, and Brenda drives at 40 km per hour. After 6 hours of driving, Heather passes a truck traveling in the opposite direction. One hour later, Brenda passes the same truck still traveling in the opposite direction. At what speed does the truck travel?**

Answer

32 km/h

Solution

Since Heather and Brenda are diving in the same direction, their relative speed is the difference of their two speeds, which is

$$52 - 40 = 12$$

km/h. Therefore, after 6 hours of driving, the distance between Heather and Brenda is

$$12 \times 6 = 72$$

km. In the hour before she passes the truck, Brenda travels 40 km. Therefore, the truck must have traveled

$$72 - 40 = 32$$

km in the same hour. Thus the truck driver drives at a speed of 32 km/h.

Problem 4.32 **David and Ray hike a mountain trail in Crystal Cove. David starts out on the trail at a pace of 4 kilometers per hour. One hour later, Ray starts out on the same trail at 6 kilometer per hour. How long will it take Ray to catch up to David?**

Answer

2 hours

Solution 1

Since David and Ray are traveling in the same direction, the relative speed is the difference of their speeds, which is

$$6 - 4 = 2$$

kilometers per hour. Since David hikes at a pace of 4 km/h, he is 4 km ahead of Ray after an hour. Thus, it will take Ray

$$4 \div 2 = 2$$

hours to catch up with David.

Solution 2

Let the time it takes Ray to catch up to David to be x hours. In those x hours, Ray travels

$$6 \times x$$

kilometers. David travels an extra hour, so he travels $x + 1$ hours. Hence he travels a total of

$$4 \times (x + 1) = 4x + 4$$

kilometers. These distances are the same, so

$$6x = 4 \times 1 + 4x$$

We solve for x to get

$$x = 2.$$

Hence, it will take Ray 2 hours to catch up to David.

Problem 4.33 **Sam begins walking at a pace of 4 km per hour from one end of a trail that is 34 km long. Ashley begins one hour later at the other end of the trail, walking towards Sam at a pace of 6 km per hour. How long will it take for them to pass each other?**

Answer

3 hours

Solution 1

In the hour before Ashley starts, Sam walks 4 km, so when Ashley starts walking they are

$$34 - 4 = 30$$

km apart. Sam and Ashley are walking toward each other, at a relative speed of

$$4 + 6 = 10$$

km/h. Therefore, it takes them

$$30 \div 10 = 3$$

hours to pass each other.

Solution 2

(Algebra) Suppose they pass each other after x hours of Ashley walking. In these x hours, Ashley walks a total of

$$6 \times x$$

km. Sam walks a total of $x + 1$ hours, so he walks a total of

$$4 \times (x + 1) = 4x + 4$$

km. When they meet they have together walked the whole trail, so

$$6x + 4x + 4 = 34$$

Solve for x to get

$$x = 3.$$

Thus they will pass each other 3 hours after Ashley begins walking.

Problem 4.34 **Tom drives his car for a round trip between place A and place B. He drives at 40 km per hour to get from A to B. At what speed should he drive back from B to A, if his average speed for the round trip is 48 km per hour?**

Answer

60 km/h

Solution

Since the distance from A to B is not given, we can assume it is a convenient number. Thus, assume the distance from A to B is 120 km. Therefore, the round trip is 240 km, so if Tom averages a speed of 48 km per hour the round trip will take

$$240 \div 48 = 5$$

hours. Similarly, the drive from A to B takes

$$120 \div 40 = 3$$

hours. Therefore the return trip must take Tom

$$5 - 3 = 2$$

hours. To travel 120 km in 2 hours, Tom must travel

$$120 \div 2 = 60$$

km per hour on the return trip.

Problem 4.35 **A boat has a rip-hole on the bottom while 20 miles away from the shore. The water comes in at a rate of 1.5 tons every minute, and the boat would sink after 70 tons of water came in. How fast must the boat go in order to reach the shore before sinking?**

Answer

$\dfrac{3}{7}$ miles per minute

Solution

We know the speed of the water comes in and the amount of water the boat can hold, then we can calculate the time,

$$70 \div 1.5 = \frac{140}{3} \text{ minutes.}$$

Then we can find the speed of the boat,

$$20 \div \frac{140}{3} = \frac{3}{7} \text{ miles/min.}$$

Therefore, the boat must go at $\frac{3}{7}$ miles per minute in order to reach the shore before sinking.

Problem 4.36 **One day, Bob rode his bike to school. When school is off, he forgot his bike and walked home instead. He spent a total of 50 minutes on the road for the round trip. If he walked for both directions, he would have spent a total of 70 minutes. How much would be the total time if he rode his bike for both directions?**

Answer

30 minutes

Solution

If it takes Bob 70 minutes to walk to and from school, his walk home from school took

$$70 \div 2 = 35$$

minutes. Therefore, the time it took for him to ride the bike to school was

$$50 - 35 = 15$$

minutes. If Bob instead had biked both ways, it would take

$$15 \times 2 = 30$$

minutes to travel to and from school.

Problem 4.37 **It takes 40 minutes for Dave to walk from home to school. It takes 15 minutes if he rides a bike instead. One day, he first rides a bike for 9 minutes before the bike breaks. He then walks the remaining distance to school. How much total time does Dave spend getting to school?**

Answer

25 minutes

Solution

Since it takes Dave a total of 15 minutes to bike to school, in 9 minutes he has biked

$$\frac{9}{15} = \frac{3}{5}$$

of the way to school. He walks the rest of the distance to school, which takes him

$$40 \times \frac{2}{5} = 16$$

minutes. Therefore, he spends a total of

$$9 + 16 = 25$$

minutes getting to school.

Problem 4.38 **Starting at the same time, a bus and a truck start traveling toward each other. After 18 hours the two vehicles meet. The bus travels at 50 miles per hour. The truck travels at 42 miles per hour, but stops for a 1 hour break after every 3 hours of travel. What is the distance between the two starting locations?**

Answer

1488 miles

Solution

We first calculate the distance the bus travels, which is

$$50 \times 18 = 900$$

miles. The truck completes a traveling and resting session every 4 hours. Since

$$18 = 4 \times 4 + 2,$$

in 18 hours the truck will actually be traveling a total of

$$3 \times 4 + 2 = 14$$

hours. Thus, the truck travels

$$42 \times 14 = 588$$

miles. Therefore, the distance between the two starting locations is

$$900 + 588 = 1488$$

miles.

Problem 4.39 **It takes a pigeon** 2 **hours to fly with the wind between two houses, and** 3 **hours against the wind. If the wind blows at a speed of** 2.5 **miles per hour, at what speed would the pigeon travel in a windless day?**

Answer

12.5 mph

Solution 1

The ratio between the time downwind and the time upwind is $2 : 3$. Because the upwind distance and downwind distance are the same, the speed ratio is $3 : 2$.

The difference between the upwind speed and downwind speed is twice the wind speed, which equals $2.5 \times 2 = 5$ mph, thus the upwind speed is 10 mph and downwind speed is 15 mph. Therefore the windless speed is $15 - 2.5 = 12.5$ mph.

Solution 2

(Algebra) The distance between the two houses is set. We use that to set up the equation. Let the speed in a windless day be x mph. When the pigeon flies with the wind, the actual speed is $x + 2.5$. When the pigeon flies against the wind, the actual speed is $x - 2.5$. The time for either way is given. Then we can set up the equation:

$$(x + 2.5) \times 2 = (x ~ 2.5) \times 3,$$

$$2x + 5 = 3x ~ 7.5,$$

$$x = 12.5.$$

Therefore the windless speed is 12.5 mph.

Problem 4.40 **Brian and David run along a circular track, starting from the same point, going opposite directions. They meet after 36 seconds. Assume that David runs the whole circle in 90 seconds. How long does it take Brian to run the whole circle?**

Answer

60 seconds

Solution 1

Since David runs the whole circle in 90 seconds, at the time they meet (36 seconds), David finishes $\frac{36}{90} = \frac{2}{5}$ of a whole circle, so Brian finishes $1 - frac25 = \frac{3}{5}$ of a whole circle. Thus the total time it takes for Brian to run the whole circle is $36 \div \frac{3}{5} = 60$ seconds.

Solution 2

(Algebra) Assume it takes Brian x seconds to run the whole circle, so he runs $\frac{1}{x}$ of a whole circle per second. Also we know that David runs $\frac{1}{90}$ of a whole circle per second. They meet after 36 seconds, so

$$\frac{1}{x} + \frac{1}{90} = \frac{1}{36},$$

$$\frac{1}{x} = \frac{1}{36} - \frac{1}{90} = \frac{1}{60},$$

therefore $x = 60$ seconds.

Problem 4.41 **An ant crawls along the sides of an equilateral triangle. It starts at one vertex and crawls at a rate per minute of 50 cm, 20 cm and 40 cm, respectively, on each of the three sides of the triangle. What is the ant's average speed as it travels around the triangle?**

Answer

$\frac{600}{19}$ cm per minute

Solution

We are not given the size of the equilateral triangle, so for ease of calculation we can assume that each side of the triangle is 200 cm long. Therefore, it takes the ant

$$200 \div 50 = 4$$

minutes to crawl along the first side,

$$200 \div 20 = 10$$

minutes for the second side, and

$$200 \div 40 = 5$$

minutes for the last side. In total, the ant crawls

$$3 \times 200 = 600$$

cm in

$$4 + 10 + 5 = 19$$

minutes. Thus, the average speed is

$$600 \div 19 = \frac{600}{19}$$

cm per minute.

Problem 4.42 A road consists of uphill, flat and downhill sections with that order. The distances of the three sections are in ratio $1 : 2 : 3$ with a total distance of 20 miles. The times JoAnn spends on the three sections are in ratio $4 : 5 : 6$. She walks at a speed of 2.5 miles per hour uphill. What is the total time she spends on the road?

Answer

5 hours

Solution

We know the ratios of the three sections and the total distance, so we can find how long each section of the road is. The uphill section is

$$20 \times \frac{1}{6} = \frac{10}{3}$$

miles long, the flat section

$$20 \times \frac{2}{6} = \frac{20}{3}$$

miles long, and the downhill section

$$20 \times \frac{3}{6} = 10$$

miles long. Since JoAnn walks 2.5 miles per hour during the uphill section, it takes her

$$\frac{10}{3} \div 2.5 = \frac{4}{3}$$

hours to do so. Since the times JoAnn spends on each section are in ratio $4 : 5 : 6$ she spends

$$\frac{4}{3} \times \frac{5}{4} = \frac{5}{3}$$

hours on the flat section and

$$\frac{4}{3} \times \frac{6}{4} = 2$$

hours on the downhill section. Therefore, the total time she spends on the road is

$$\frac{4}{3} + \frac{5}{3} + 2 = 5$$

hours.

Problem 4.43 **A bridge consists of three sections of equal length: an uphill section, a flat section and a downhill section. At what average speed does a cyclist ride his bicycle if he travels the three sections at a speed of 4 meters per second, 6 meters per second and 8 meters per second respectively?**

Answer

$\frac{72}{13}$ meters per second

Solution

The distance of each section is not given in the problem, so we can use this to our advantage and assume each section is 240 meters long. With this assumption, the cyclist takes

$$240 \div 4 = 60$$

seconds on the uphill section,

$$240 \div 6 = 40$$

seconds on the flat section, and

$$240 \div 8 = 30$$

seconds on the downhill section. The total distance of the bridge is

$$3 \times 240 = 720$$

meters and the total time is

$$60 + 40 + 30 = 130$$

seconds. Therefore, the cyclist's average speed is

$$720 \div 130 = \frac{72}{13} \approx 5.54$$

meters/sec.

Problem 4.44 At 6 AM, bus station A starts to dispatch buses to station B, and station B starts to dispatch buses to station A. They each dispatch one bus to the other station every 8 minutes. The one-way trip takes 45 minutes. One passenger gets on the bus at station A at 6:16 AM. How many buses coming from station B will the passenger see en route?

Answer

8

Solution

The passenger gets on the bus at 6:16 AM to travel to station B. Since the one-way trip takes 45 minutes, none of the buses from station B have arrived at station A when the passenger departs. The passenger will therefore see every bus that departs station B between 6 AM and their arrival time at station B which is 7:01 AM. One bus leaves exactly at 6 AM, and in the 61 minutes that follow 7 more buses leave, because

$$61 \div 8 \approx 7.625$$

and we round down because the buses leave at the end of each 8 minute interval. Therefore, the passenger sees a total of

$$7 + 1 = 8$$

buses en route from station A to station B.

Problem 4.45 **Sam walks up a hill. After every 30 minutes of walking he takes 10 minutes to rest. When he walks down the hill, he instead rests for 5 minutes after every 30 minutes of walking. Sam walks downhill 1.5 times faster than he walks uphill. If he spends 3 hours and 50 minutes traveling up the hill, how much time does he spend traveling down the hill?**

Answer

2 hours 15 minutes

Solution

We know it takes 3 hours and 50 minutes, or

$$3 \times 60 + 50 = 230$$

minutes to travel uphill. We want to find how much of this time Sam was actually walking. A walking/resting cycles takes 40 minutes, so because

$$230 = 5 \times 40 + 30$$

the 230 minutes will consist of 5 walking/resting cycles and end with Sam walking the final 30 minutes. Hence, he walks a total of

$$5 \times 30 + 30 = 180$$

minutes. Since Sam walks downhill at a speed 1.5 times faster as that he walks uphill, he take will spend 1.5 times less time walking downhill, a total of

$$180 \div 1.5 = 120$$

minutes. In these 120 minutes, me must rest

$$(120 \div 30) - 1 = 3$$

times (since he can walk the final 30 minutes). He therefore spends a total of

$$120 + 3 \times 5 = 135$$

minutes, or 2 hours and 15 minutes to travel uphill.

Problem 4.46 **Omar walks up a hill. After every 40 minutes of walking uphill he takes 10 minutes to rest. Downhill he rests for 5 minutes after every 40 minutes of walking. Omar walks downhill at a speed 2 times as fast as that he walks uphill. If he spends 2 hours traveling down the hill, how much time does he spend traveling up the hill?**

Answer

4 hours and 30 minutes

Solution

It takes 2 hours or 120 minutes to travel down the hill. Note that

$$120 = 45 + 5 + 45 + 5 + 20$$

so Omar rests for 10 total minutes and therefore spends 110 minutes walking down the hill. Since Omar walks downhill twice as fast as up hill, he will take twice the time to walk uphill. Therefore Omar spends

$$2 \times 110 = 220$$

minutes actually walking up the hill. Omar must rest every 40 minutes, so because

$$220 \div 40 = 5.5$$

Omar must rest 5 times, a total of 50 minutes. Thus Omar spends

$$220 + 50 = 270$$

minutes or 4 hours and 30 minutes traveling uphill.

Problem 4.47 **It takes a ship 6 hours to travel downstream between two piers, and 8 hours upstream. If the water flows at a speed of 2.5 miles per hour, at what speed would the ship travel in still water?**

Answer

17.5 miles per hour

Solution 1

Consider the same ship traveling upstream and downstream. The flow of the water adds 2.5 miles per hour upstream and subtracts 2.5 miles per hour downstream. Therefore the difference in speeds between the boat traveling upstream and downstream is 5 miles per hour. Note that in 24 hours of traveling both upstream and downstream, the ship can make

$$24 \div 6 = 4$$

trips between the piers traveling downstream and

$$24 \div 8 = 3$$

trips between the piers traveling upstream. Since the ship is traveling the same amount of time in both directions, the difference in the distance it travels must be due to the 5 mile per hour difference in speed due to the water. Hence, the difference between the two piers is

$$5 \times 24 = 120$$

miles. Since it takes 6 hours for the to travel downstream between the piers, it travels at a speed of
$$120 \div 6 = 15$$

miles per hour. Therefore, in still water it travels at a speed of

$$15 + 2.5 = 17.5$$

miles per hour.

Solution 2

(Ratios) The flow of the water adds 2.5 miles per hour upstream and subtracts 2.5 miles per hour downstream. Therefore the difference in speeds between the boat traveling upstream and downstream is 5 miles per hour. The distance between the piers is the same whether we travel upstream or downstream. Therefore, because we know the ratio of times between the boat traveling upstream and downstream from pier to pier is

$$8 : 6 = 4 : 3,$$

the speed of the boat traveling upstream and downstream is in the opposite ratio

$$3 : 4.$$

We want the speeds to differ by 5 miles per hour, so multiplying both sides of the ratio by 5 we have the speeds are

$$15 : 20$$

so the speed upstream is 15 miles per hour and the speed downstream is 20 miles per hour. Therefore, the speed in still water is

$$15 + 2.5 = 20 - 2.5 = 17.5$$

miles per hour.

Solution 3

Let the speed of the ship would travel in still water be x miles per hour. Since the water flows at a rate of 2.5 miles per hour, the speed upstream is

$$x - 2.5$$

miles per hour and the speed downstream is

$$x + 2.5$$

miles per hour. The distance between the piers is the same upstream or downstream, so since it takes 8 hours upstream and 6 hours downstream,

$$(x - 2.5) \times 8 = (x + 2.5) \times 6.$$

We can then solve to get,

$$x = 17.5,$$

so the ship travels at a speed of 17.5 miles per hour in still water.

Problem 4.48 George goes to school by riding his bike to the bus station, taking the bus, and then walking to his classroom. The ratio of the three distances is 2:8:1. His biking speed is 10 mph. The bus travels at a speed of 50 mph. His walking speed is 2 mph. What is his average speed in his way to school?

Answer

$$\frac{550}{43} \text{ mph}$$

Solution

We only are given a ratio for the three distances, so we can assume the distance George bikes, buses, and walks are 2 miles, 8 miles, and 1 mile respectively. Then the total distance is

$$2 + 8 + 1$$

miles. Next we calculate the total time spent. The biking time is

$$2 \div 10 = \frac{1}{5}$$

hours; the bus time is

$$8 \div 50 = \frac{4}{25}$$

hours; the walking time is

$$1 \div 2 = \frac{1}{2}$$

hours. Thus the total time is

$$\frac{1}{5} + \frac{4}{25} + \frac{1}{2} = \frac{43}{50}$$

hours. We can then calculate the average speed, which is

$$11 \div \frac{43}{50} = \frac{550}{43} \approx 12.79$$

mph.

Problem 4.49 Cindy rides her bike from home to school at a speed that is 120 meters per minute faster than if she walks, and the time she spends is 3/5 less than if she walks. How fast does Cindy walk from home to school?

Answer

80 meters per minute

Solution 1

We are not given a distance or time in the problem, so we may assume for convenience that Cindy walks to school for 5 minutes. Since the time she spends biking is $\frac{3}{5}$ less than the time she spends walking, she spends

$$5 \times \frac{2}{5} = 2$$

minutes biking to school. Since Cindy bikes 120 meters per minute faster than she walks (and can bike to school in 2 minutes), if Cindy walks for 2 minutes, she will have

$$2 \times 120 = 240$$

meters left to walk before she gets to school. Since it takes her 5 minutes in total to talk to school, she can walk 240 meters in

$$5 - 2 = 3$$

minutes. Hence, Cindy can walk

$$240 \div 3 = 80$$

meters per minute.

Solution 2

(Algebra) Let the speed that Cindy walks to be x meters per minute. Since the time she spends biking is $\frac{3}{5}$ less than the time she spends walking, she bikes for $\frac{2}{5}$ of the time she bikes. Since the distances she travels are the same, we have

$$x = \frac{2}{5} \times (x + 120).$$

Distributing and grouping like terms we have

$$\frac{3}{5}x = 48.$$

Then we can solve for x,

$$x = 48 \times \frac{5}{3} = 80.$$

Therefore, Cindy walks at a speed of 80 meters per minute.

Problem 4.50 **Joe and JoAnn walk toward each other from two locations that are 36 miles apart. If Joe departed 2 hours earlier, they would meet 2.5 hours after JoAnn departed. If JoAnn departed 2 hours earlier, they would meet 3 hours after Joe departed. Find the respective speed at which each walks.**

Answer

Joe: 6 miles/hr, JoAnn: 3.6 miles/hr

Solution 1

Since the two locations are 36 miles apart, when Joe and JoAnn meet they have walked a combined 36 miles. In the first scenario, Joe walks a total of 4.5 hours and JoAnn walks a total of 2.5 hours. In the second scenario, Joe walks a total of 3 hours and JoAnn walks a total of 5 hours. If we double the first scenario, we see that if Joe walks

$$4.5 \times 2 = 9$$

hours and JoAnn walks for

$$2.5 \times 2 = 5$$

hours, they walk a combined

$$36 \times 2 = 72$$

miles. Comparing this with the second scenario (since JoAnn walks 5 hours in each case), we see that in

$$9 - 3 = 6$$

hours, Joe must walk

$$72 - 36 = 36$$

miles by himself. Therefore Joe walks

$$36 \div 6 = 6$$

miles/hr. Thus, in 3 hours, Joe walks

$$6 \times 3 = 18$$

miles, so using the second scenario JoAnn must walk the remaining

$$36 - 18 = 18$$

miles in 5 hours. Thus, JoAnn walks at a speed of

$$18 \div 5 = 3.6$$

miles/hr.

Solution 2

Let's assume the Joe's walking speed is x miles/hr, and JoAnn's walking speed is y miles/hr. If they meet 2.5 hours after JoAnn departs, JoAnn walks for a total of 2.5 hours, while Joe walks 2 extra hours, for a total of 4.5 hours. This gives us the equation

$$4.5 \times x + 2.5 \times y = 36.$$

If they meet 3 hours after Joe departs, Joe walks a total of 3 hours. JoAnn walks 2 extra hours, so she walks 5 hours. This gives another equation

$$3 \times x + 5 \times y = 36.$$

This gives us the system of equations

$$\begin{cases} 4.5x + 2.5y &= 36, \\ 3x + 5y &= 36. \end{cases}$$

Multiplying the first equation by 2 we get

$$9x + 5y = 72.$$

If we subtract the second equation from this we have

$$6x = 36,$$

so

$$x = 6.$$

Substituting back into the second equation we have

$$3 \times 6 + 5y = 36$$

so combining like terms we have

$$5y = 18$$

and solving for y gives

$$y = \frac{18}{5} = 3.6$$

Therefore, Joe walks at a speed of 6 miles/hr, and JoAnn walks at a speed of 3.6 miles/hr.

Problem 4.51 **Sami and Rajan practice running together. If Sami starts to run after Rajan runs for 10 meters, then it will take Sami 5 seconds to catch up with Rajan. If Sami starts to run after Rajan runs for 2 seconds, then it will take Sami 4 seconds to catch up with Rajan. How fast can each person run?**

Answer

Sami: 6 meters/sec, Rajan: 4 meters/sec

Solution 1

In the second situation, Sami and Rajan run the same distance if Sami runs for 4 seconds and Rajan runs for 6 seconds. Thus, the speed at which they run is in the opposite ratio

$$6 : 4 = 3 : 2.$$

This also tells us that if they run the same amount of time, the ratio of the distances Sami and Rajan run is also

$$3 : 2.$$

In the first scenario, Sami needs to run 10 extra meters and Rajan (both running for 5 seconds). Multiplying the above ratio by 10 we get

$$30 : 20$$

so Sami runs for 30 meters and Rajan runs 20 meters. Since this takes them each 5 seconds, we see that Sami runs at a rate of

$$30 \div 5 = 6$$

meters/sec and Rajan runs at a rate of

$$20 \div 5 = 4$$

meters/sec.

Solution 2

(Algebra) Let's assume the Sami's running speed is x meters/sec, and Rajan's running speed is y meters/sec. In the first situation, Sami runs a distance of $5 \times x$, while Rajan runs a distance of $10 + 5 \times y$, so

$$5 \times x = 10 + 5 \times y.$$

In the second situation, Sami runs for 4 seconds, while Rajan runs for 6 seconds, so

$$4 \times x = 6 \times y.$$

This gives us the system of equations

$$\begin{cases} 5x &=& 5y + 10, \\ 4x &=& 6y. \end{cases}$$

Dividing the first equation by 5 gives us

$$x = y + 2$$

so we can substitute this into the second equation to get

$$4 \times (y + 2) = 6y.$$

Distributing and combining like terms we have

$$8 = 2y$$

so we can solve for y and get

$$y = 4.$$

Substituting back into the second equation we have

$$4x = 6 \times 4$$

so we can solve for x:

$$x = 6.$$

Therefore, Sami runs at a speed of 6 meters/sec, and Rajan runs at a speed of 4 meters/sec.

Problem 4.52 It takes 25 seconds for a train to pass completely pass through a tunnel which measures 250 meters long. It takes 23 seconds for the train to completely pass through another tunnel which measures 210 meters long. How long does it take the train to pass an approaching train which is 320 meters long and at the speed of 18 m/s?

Answer

15 seconds

Solution

We first calculate the speed of the original train. Note to completely pass through the tunnel the train must travel the length of the tunnel plus the length of the train. Since the length of the train does not change and it takes 25 seconds to pass through a 250 meter tunnel and 23 seconds to pass through a 210 meter tunnel it must travel the extra

$$250 - 210 = 40$$

meters in

$$25 - 23 = 2$$

seconds. Therefore the speed of the train is

$$40 \div 2 = 20$$

m/s. Therefore, in the 25 seconds it takes to pass through the 250 meter tunnel, the train travels

$$20 \times 25 = 500$$

meters. Since the tunnel itself is 250 meters long, the difference

$$500 - 250 = 250$$

meters must be the length of the original train.

We can now find how long it takes to pass the approaching train. To pass each other, the trains must travel a distance of

$$320 + 250 = 570$$

meters (the sum of the lengths of both trains). Since the trains are approaching each other, their relative speed is

$$18 + 20 = 38$$

m/s. Therefore, it takes

$$570 \div 38 = 15$$

seconds for the original train to pass the approaching train.

Problem 4.53   **Ming and Ping both take a walk. The distance that Ming walks is 1/5 less than the distance that Ping walks. Also, the time Ping spends is 1/8 more than the time Ming spends. What is the ratio of their respective speed?**

Answer

$9 : 10$.

Solution

We are not given specific distances or times, so we pick convenience numbers. Suppose Ping walks 10 miles. Then since Ming walks $\frac{1}{5}$ less distance than Ping, they walk

$$10 \times \frac{4}{5} = 8$$

miles. Similarly, suppose Ming spends 8 hours walking. Then since Ping spends $\frac{1}{8}$ more time, they spend

$$8 \times \frac{9}{8} = 9$$

hours walking. Therefore Ming walks at a rate of

$$8 \div 8 = 1$$

mile per hour, while Ping walks at a rate of

$$10 \div 9 = \frac{10}{9}.$$

Therefore the ratio between their respective speeds is

$$1 : \frac{10}{9} = 9 : 10$$

as asked by the problem.

Problem 4.54 **Tyler and Hannah start to walk on the same direction from the same place. Tyler walks at 5 miles per hour. Hannah walks at 1 mile per hour for the first hour, 2 miles per hour for the second hour. Hannah increases her speed by 1 mile per hour after each hour. How long does it take for Hannah to catch up with Tyler?**

Answer

9 hours

Solution 1

Tyler travels 5 miles every hour, so he travels $5, 10, 15, \ldots$ miles after $1, 2, 3, \ldots$ hours. Hannah increases her speed each hour, so she travels $1, 1 + 2 = 3, 1 + 2 + 3 = 6, \ldots$ miles after $1, 2, 3, \ldots$ hours. We can continue in this manner to find when Hannah catches up to Tyler. The results are summarized in the table below:

Hours	1	2	3	4	5	6	7	8	9
Tyler's distance	5	10	15	20	25	30	35	40	45
Hannah's distance	1	3	6	10	15	21	28	36	45

Therefore, it takes Hannah 9 hours to catch up with Tyler.

Solution 2

Tyler travels 5 miles per hour every hour. Hannah starts at 1 mile per hour and increases by 1 every hour. Therefore, they are both traveling at the same speed in hour 5. Since Hannah keeps increasing her speed, every hour after 5 makes up the ground she lost in an hour before 5. For example, in hour 6 she travels $6 = 5 + 1$ miles making up for the $4 = 5 - 1$ miles she traveled in hour 4. Therefore, in

$$5 - 1 = 4$$

more hours, Hannah will catch up to Tyler. This takes a total of

$$5 + 4 = 9$$

hours.

Problem 4.55   **A boat takes 3 days to travel from town A to town B, but it takes 4 days to travel from town B to town A. If a motor-less raft is left alone in the water by town A, how long will it take for the raft to float to town B?**

Answer

24 days

Solution 1

Consider the boat traveling 12 days downstream and 12 days upstream. 12 days downstream is enough for

$$12 \div 3 = 4$$

trips from town A to town B (downstream) and similarly 12 days upstream is enough for

$$12 \div 4 = 3$$

trips from town B to town A (upstream). Since the boat travels the same amount of time upstream and downstream, the extra trip downstream is because of the water flow. Therefore, a motor-less raft floating downstream from town A will take

$$12 + 12 = 24$$

days to reach town B.

Solution 2

(Algebra) Let the speed of the water flow to be w and the speed of the boat to be b. Then the downstream speed is $b + w$ and the upstream speed is $b - w$. Since it takes 3 days to travel (downstream) from town A to town B but 4 days to make the same trip back (upstream) we can set up the following equation,

$$(b + w) \times 3 = (b - w) \times 4.$$

Distributing and combining like terms we have

$$b = 7w,$$

so the downstream speed is

$$b + w = 7w + w = 8w.$$

Now let the time it takes for the raft to float to town B to be r days, then we have the following equation,

$$8w \times 3 = w \times r.$$

The variable w cancels from both sides, and solving for r, we have

$$r = 24.$$

Hence it will take the motor-less raft 24 days to float to town B.

Problem 4.56 **Starting at the same time, Cathy and David drive two cars toward each other from the two ends, call them A and B, of the same road. Cathy drives 1.2 times faster than David. When they pass by each other, they are 8 miles away from the halfway point between A and B. Find the total length of the road.**

Answer

176 miles

Solution 1

Cathy drives 1.2 times faster than David, so since

$$1.2 = \frac{6}{5}$$

the ratio of their speeds is

$$6 : 5.$$

Since they both travel the same amount of time, this means the ratio between the distance Cathy travels and the distance David travels is also 6 : 5. We know that when they pass each other they are 8 miles from the halfway point, meaning that Cathy must have driven 16 miles more than David. Multiplying both sides of the ratio 6 : 5 by 16 we see that

$$6 : 5 = 96 : 80,$$

so Cathy must drive 96 miles and David must drive 80 miles. Therefore the total length of the road (which combined they have driven when they meet) is

$$96 + 80 = 176$$

miles.

Solution 2

(Algebra) Cathy drives 1.2 times faster than David, so since

$$1.2 = \frac{6}{5}$$

the ratio of their speeds is

$$6 : 5.$$

Since Cathy and David both travel the same about of time, the ratio between their distances is also 6 : 5. Let x be such that Cathy travels $6x$ miles and David travels $5x$ miles. They pass each other 8 miles past the halfway's point, meaning that Cathy drives 16 miles more than David. Thus,

$$6x = 5x + 16$$

so combining like terms we have

$$x = 16$$

miles. Therefore the length of the road is

$$6x + 5x = 11x = 11 \times 16 = 176$$

miles.

Problem 4.57 **Alice, Bob, and Cindy drive their cars separately from site A to site B simultaneously. Alice drives at 60 mph and Bob drives at 48 mph. Alice passes a car from the opposite direction after 6 hours of driving. One hour later, Bob pass the same car still traveling in the opposite direction. One more hour later, Cindy also passes the same car. Find the speed at which Cindy drives her car.**

Answer

39 mph

Solution

Alice travels 60 mph, so she passes the car after she travels

$$60 \times 6 = 360$$

miles. Similarly, Bob passes the car after traveling

$$48 \times 7 = 336$$

miles. Therefore, the car moving the opposite direction traveled

$$360 - 336 = 24$$

miles in one hour, hence is traveling 24 mph. Therefore, after another hour, the car traveling the opposite direction will be

$$336 - 24 = 312$$

miles from site A when Cindy passes it. Since Cindy has been traveling 8 hours when this happens, Cindy's speed is

$$312 \div 8 = 39$$

mph.

Problem 4.58 **Amy and Amanda walk on a circular track. Amy starts from spot A, and Amanda starts from spot B. One walks clockwise and the other counterclockwise. After 6 minutes, they meet. Four minutes later, Amy arrives at spot B. After 8 more minutes of walking, they meet again. How many minutes does it take for each to walk one full circle?**

Answer

Amy: 20 minutes, Amanda: 30 minutes

Solution

Amanda walks from spot B to the first meeting point in the first 6 minutes. It takes Amy 4 minutes to walk from the first meeting point to B (the same distance but in the opposite direction). These times for Amy and Amanda are in ratio

$$4 : 6 = 2 : 3,$$

meaning the speed at which they walk is in the opposite ratio $3 : 2$. Amy and Amanda are pass each other at the first meeting point, and meet again at the second meeting point

$$4 + 8 = 12$$

minutes later, at which point they have together walked a full circle. Since Amy walks $\frac{3}{2}$ the speed of Amanda, Amy can walk the other portion of the track in

$$12 \div \frac{3}{2} = 8$$

minutes, meaning she can complete one full circle in

$$12 + 8 = 20$$

minutes. Since Amanda walks $\frac{2}{3}$ the speed, it takes her

$$20 \div \frac{2}{3} = 30$$

minutes to complete one full circle.

Problem 4.59 **A railroad bridge measures 1000 meters long. A train passes the bridge. It takes 120 seconds from the time the train enters the bridge to the time the whole train gets off the bridge. There are 80 seconds during which time the whole train is on the bridge. Find both the speed and the length of the train.**

Answer

Speed: 10 meters/sec. Length: 200 meters

Solution 1

Since the whole train is on the bridge for 80 seconds, it takes

$$120 - 80 = 40 \text{ seconds}$$

for the train to move on and off the bridge. Since the train takes the same amount of time to move on the bridge and move off the bridge it takes

$$40 \div 2 = 20 \text{ seconds}$$

to do each. If we focus on the head of the train, it takes

$$80 + 20 = 100 \text{ seconds}$$

for the head of the train to travel the 1000 meter length of the bridge. Hence the train moves at a speed of

$$1000 \div 100 = 10 \text{ meters/sec.}$$

Finally, since it takes 20 seconds to move off the bridge we know the length of the train is

$$20 \times 10 = 200 \text{ meters.}$$

Solution 2

(Algebra) Let the length of the train be x meters. The distance the train travels in the 120 seconds from entering the bridge to fully leaving the bridge is $1000 + x$ so the trains speed is

$$(1000 + x) \div 120 = \frac{1000 + x}{120}$$

meters/sec. However, we also know the whole train is on the track for 80 seconds, in which it travels $1000 - x$ meters, so the speed is also

$$(1000 - x) \div 80 = \frac{1000 - x}{80}$$

meters/sec. Therefore, we can set up the following equation,

$$\frac{1000 + x}{120} = \frac{1000 - x}{80}.$$

Clearing denominators we get

$$2000 + 2x = 3000 - 3x$$

and grouping like terms gives us

$$5x = 1000$$

so we can solve for x:

$$x = 1000 \div 5 = 200.$$

Then the speed of the train is

$$\frac{1000 + 200}{120} = 10$$

meters/sec. Therefore, the speed of the train is 10 m/s and the length of the train is 200 meters.

Problem 4.60 **A hunting dog chases a hare 21 meters ahead. The dog runs in a series of jumps, with each jump being 3 meters long. Each jump for the hare is 2.1 meters. If the dog jumps three times for every four times the hare jumps, how much farther can the hare travel before the dog catches it?**

Answer

294 meters

Solution 1

Each jump for the dog is 3 meters, so in 3 jumps the dog moves a total of

$$3 \times 3 = 9$$

meters. In the same time the hare jumps 4 times, for a total of

$$4 \times 2.1 = 8.4$$

meters. Therefore, every time the hare jumps 4 times, the dog catches up

$$9 - 8.4 = 0.6$$

meters. Since

$$21 \div 0.6 = 21 \div 3/5 = 35$$

the hare can jump a total of

$$35 \times 4 = 140$$

times before the dog catches up. Thus, the hare travels a total of

$$140 \times 2.1 = 294$$

meters before the dog catches it.

Solution 2

(Algebra) Since the dog jumps three times for every four times the hare jumps, the ratio of jumps is $3 : 4$. Let x be such that the dog jumps $3x$ times and the hare jumps $4x$ times when the dog catches the hare. In that time, the dog will travel

$$3 \times 3x = 9x$$

meters. The hare starts 21 meters ahead and jumps a total of

$$2.1 \times 4x = 8.4x$$

meters. Since the dog catches up to the hare,

$$9x = 20 + 8.4x$$

Combining like terms we have

$$0.6x = 21$$

so we can solve for x, to get

$$x = 21 \div 0.6 = 21 \div \frac{3}{5} = 35.$$

Therefore the hare can travel

$$2.1 \times 4 \times 35 = 294$$

more meters before the dog catches it.

5. Work Related Problems

Problem 5.1 Eldridge can split a cord of wood in 4 days and his father can do it in 3 days. How long would it take them if they worked together?

Answer

$\dfrac{12}{7}$ days

Solution

Each day Eldridge can split $\dfrac{1}{4}$ of the wood, and his father can split $\frac{1}{3}$ of the wood. Together they can split

$$\frac{1}{4} + \frac{1}{3} = \frac{7}{12}$$

of the wood per day. So it takes them $\dfrac{12}{7}$ days if they worked together.

Problem 5.2 Using a ride-on lawn mower, Abby can mow the lawn in 2 hours.

Her sister Carla takes 3 hours using an older mower. How long will it take them if they work together?

Answer

$\dfrac{6}{5}$ hours

Solution

Each hour Abby can mow $\dfrac{1}{2}$ of the lawn, and Carla can mow $\dfrac{1}{3}$ of the lawn. Together they can mow

$$\frac{1}{2} + \frac{1}{3} = \frac{5}{6}$$

of the lawn per hour. Thus it takes $\dfrac{6}{5}$ hours for them to mow the lawn together.

Problem 5.3 **One drain pipe can empty a swimming pool in 6 hours. Another pipe takes 3 hours. If both pipes are used simultaneously to drain the pool, how long does it take the drain the pool?**

Answer

2 hours

Solution

The first pipe can empty $\dfrac{1}{6}$ of the pool per hour, and another pipe can empty $\dfrac{1}{3}$ of the pool per hour. Together they can empty

$$\frac{1}{6} + \frac{1}{3} = \frac{1}{2}$$

of the pool per hour. Hence, it takes 2 hours for them to empty the pool together.

Problem 5.4 **Tom and Jerry can finish organizing the books at school's library together in 5 hours. If Tom do it alone, it will take him 8 hours. How long would it take Jerry to finish the same task alone?**

$\frac{40}{3}$ hours

Solution

Tom can organize $\frac{1}{8}$ of the book per hour. When working together, they can organize $\frac{1}{5}$ of the book per hour. Therefore, we can find the amount of work Jerry can do alone per hour

$$\frac{1}{5} - \frac{1}{8} = \frac{3}{40}.$$

Since Jerry can organize $\frac{3}{40}$ of the book per hour, it would take him $\frac{40}{3}$ hours to finish the same task alone.

Problem 5.5 **Stan can load the truck in 40 minutes. If I help him, it takes us 15 minutes. How long will it take me alone?**

Answer

24 minutes

Solution

Stan can load $\frac{1}{40}$ of the truck per minute. When we work together, we can load $\frac{1}{15}$ of the truck per minute. Then, we can find the amount of work that I can do per minute

$$\frac{1}{15} - \frac{1}{40} = \frac{1}{24}.$$

Since I can load $\frac{1}{24}$ of the truck per minute, it would take me 24 minutes to finish it alone.

Problem 5.6 Suppose Chris can paint the entire house in fourteen hours, and Bill can do it in ten hours. How long would it take the two painters working together to paint the house?

Answer

$$\frac{35}{6}$$

Solution

Chris can paint $\frac{1}{14}$ of the house per hour, and Bill can paint $\frac{1}{10}$ of the house per hour. Together they can paint

$$\frac{1}{14} + \frac{1}{10} = \frac{6}{35}$$

of the house per hour. Therefore, it takes $\frac{35}{6}$ hours for them to paint the house together.

Problem 5.7 Two mechanics in the maintenance department were working on Daniel's car. One can complete the maintenance service in 5 hours, but the other mechanic, who is new, needs 8 hours. How long would it take the two mechanics working together to finish the service?

Answer

$$\frac{40}{13} \text{ hours}$$

Solution

The first mechanic can complete $\frac{1}{5}$ of the maintenance service per hour, and the new mechanic can complete $\frac{1}{8}$ of the maintenance service per hour. Together they can complete

$$\frac{1}{5} + \frac{1}{8} = \frac{13}{40}$$

of the maintenance service per hour. Therefore, it takes $\dfrac{40}{13}$ hours for them to complete the maintenance service together.

Problem 5.8 **Billy and Tim can paint a fence in 4 hours together. It is known that Billy can paint the same fence alone in 6 hours. How long would it take Tim to paint the fence alone?**

Answer

12 hours

Solution

Billy can paint $\dfrac{1}{6}$ of the fence per hour. When Billy and Tim work together, we can paint $\dfrac{1}{4}$ of the fence per hour. Then, we can find the amount of work that Tim can do per hour

$$\frac{1}{4} - \frac{1}{6} = \frac{1}{12}.$$

Since Tim can paint $\dfrac{1}{12}$ of the fence per hour, it would take Tim 12 hours to paint the fence alone.

Problem 5.9 **Kathy takes 3 hours to wash 300 dishes, and Andrew takes 2.5 hours to wash 300 dishes. How long will they take together to wash 1100 dishes?**

Answer

5 hours

Solution

In one hour, Kathy can wash $300 \div 3 = 100$ dishes, and Andrew can wash $300 \div 2.5 = 120$ dishes.

If they work together, in one hour they can wash

$$100 + 120 = 220 \text{ dishes.}$$

Therefore, it will take them

$$1100 \div 220 = 5 \text{ hours}$$

to finish washing 1100 dishes together.

Problem 5.10 **Amy and her sister Clair's house has a 420 square foot lawn in the back. Amy can mow 120 square feet in 30 minutes. When Amy and Clair work together, they can finish the whole lawn in one hour. How many square feet per minute can Clair mow?**

Answer

3

Solution

Amy can mow

$$120 \div 30 = 4$$

square feet per minute of the lawn. When Amy and Clair work together, they can finish mowing the lawn in one hour. This means they can mow

$$420 \div 60 = 7$$

square feet per minute. Then, we can find the amount of work that Clair can do per minute

$$7 - 4 = 3.$$

Therefore, Clair can mow 3 square feet per minute.

Problem 5.11 **If 8 people can finish a task in thirteen days, how many days would it take to do the same job with 16 people?**

Answer

6.5 days

Solution

If 8 people can finish the task in thirteen days, then when we double the amount of people, it will cut the time it takes to do the same job in half. Therefore, it would take 6.5 days for 16 people to do the same job.

Problem 5.12 **It takes 1.5 hours for Jim to water the plants in a garden. Lily can water the same amount of plants in 2 hours. How long will it take Jim and Lily, work together to water the plants in the garden?**

Answer

$\dfrac{6}{7}$ hours

Solution

Jim can water

$$\frac{1}{1.5} = \frac{2}{3}$$

of the plants per hour, and Lily can water $\dfrac{1}{2}$ of the plants per hour. Together they can water

$$\frac{2}{3} + \frac{1}{2} = \frac{7}{6}$$

of the plants per hour.

So, it takes $\dfrac{6}{7}$ hours for them to water the plants together.

Problem 5.13 **It takes pump A 6 hours to fill the pool, pump B 8 hours, and pump C 4.8 hours. How long would it take the three pumps together to fill the pool?**

Answer

2 hours

Solution

Pump A can fill $\frac{1}{6}$ of the pool per hour, pump B can fill $\frac{1}{8}$ of the pool per hour, and pump C can fill

$$\frac{1}{4.8} = \frac{5}{24}$$

of the pool per hour. Together they can fill

$$\frac{1}{6} + \frac{1}{8} + \frac{5}{24} = \frac{1}{2}$$

of the pool per hour. It thus takes 2 hours for all three pumps together to fill the pool.

Problem 5.14 **Nancy can row a boat across the river in 45 minutes, while Susan can do it in 35. If both of them sit in one boat and row together, how long will it take?**

Answer

$\frac{315}{16}$ minutes

Solution

Nancy can row $\frac{1}{45}$ of the width of the river per minute, and Susan can row $\frac{1}{35}$ of the width of the river per minute. Together they can row

$$\frac{1}{45} + \frac{1}{35} = \frac{16}{315}$$

of the width of the river per minute.

So, it takes $\frac{315}{16}$ minutes for them to across the river together.

Problem 5.15 **Employee A can complete a task in 3 hours. When working with Employee B, they can complete it in 2 hours. How long does it take for Employee B to finish the task if he/she works alone?**

Answer

6 hours

Solution

Employee A can complete $\frac{1}{3}$ of the task per hour. When Employee A and Employee B work together, they can complete $\frac{1}{2}$ of the task per hour. Then, we can find the amount of work that Employee B can do per hour

$$\frac{1}{2} - \frac{1}{3} = \frac{1}{6}.$$

Since Employee B can complete $\frac{1}{6}$ of the task per hour, it takes Employee B 6 hours to complete the task alone.

Problem 5.16 **Niki always leaves her cell phone on. If her cell phone is on but she is not actually using it, the battery will last for 24 hours. If she is using it constantly, the battery will last for only 3 hours. Since the last recharge, her phone has been on 9 hours, and during that time she has used it for 60 minutes. If she doesn't talk any more (but leaves the phone on), how many more hours will the battery last?**

Answer

8 hours

Solution 1

If the phone is on but not in use, it uses $\frac{1}{24}$ of the battery per hour. If it is in use it uses $\frac{1}{3}$ of the battery per hour. Since the phone has been on for 9 hours, during which it is in use for 1 hour, and not in use for 8 hours, Niki has used

$$8 \times \frac{1}{24} + 1 \times \frac{1}{3} = \frac{2}{3}$$

of the battery. She therefore has $\frac{2}{3}$ of the battery left, which will last

$$\frac{2}{3} \div \frac{1}{24} = 8$$

more hours if the phone is on but not in use.

Solution 2

Since the battery can last for 3 hours if she uses it constantly and 24 hours without using it, using the phone for 1 hour is equivalent to 8 hours without using it. Therefore, as the phone has been on but not in use for $9 - 1 = 8$ hours and in use for 1 hour, we can view the phone as being on for $8 + 8 = 16$ hours. If she doesn't talk any more the battery will last

$$24 - 16 = 8$$

more hours.

Problem 5.17   **There is a 10,000 liter swimming pool in Lance's community. The pool has two pipes: A and B. Pipe B delivers 1,000 liters water per hour. When pipe A and B are both on, the pool can be filled in 4 hours. How many liters per hour can pipe A deliver?**

Answer

1500 liters

Solution

When pipe A and pipe B are both on, they can fill the pool in 4 hours, which means they deliver

$$10000 \div 4 = 2500$$

liters of water per hour. Then, we can find the amount of water that pipe A can deliver per hour

$$2500 - 1000 = 1500 \text{ liters}.$$

Problem 5.18   **Bob, John, and Calvin can paint a wall alone in 2 hours, 2.5 hours, and 1.5 hours, respectively. How long does it take if all three of them work together?**

Answer

$\dfrac{30}{47}$ hours.

Solution

Bob can paint $\frac{1}{2}$ of the wall per hour, John can paint

$$\frac{1}{2.5} = \frac{2}{5}$$

of the wall per hour, and Calvin can paint

$$\frac{1}{1.5} = \frac{2}{3}$$

of the wall per hour. Together they can paint

$$\frac{1}{2} + \frac{2}{5} + \frac{2}{3} = \frac{47}{30}$$

of the wall per hour. Thus, it takes $\frac{30}{47}$ hours for Bob, John, and Calvin together to paint the wall.

Problem 5.19 **Joe's mom can clean the kitchen in 45 minutes. If Joe helps his mother, they can clean it in 30 minutes. How long would it take Joe to clean it by himself?**

Answer

90 minutes

Solution

Joe's mom can clean $\frac{1}{45}$ of the kitchen per minute. When Joe and his mom work together, they can clean $\frac{1}{30}$ of the kitchen per minute. Then, we can find the amount of work that Joe can do per hour

$$\frac{1}{30} - \frac{1}{45} = \frac{1}{90}.$$

Since Joe can clean $\frac{1}{90}$ of the kitchen per minute, it would take Joe 90 minutes to clean the kitchen alone.

Problem 5.20 **Melisa can finish a project in 2 hours, while Terry works 1.5 times faster than Melisa. How long would it take them to finish together?**

Answer

$\dfrac{4}{5}$ hours

Solution

Terry works 1.5 times faster than Melisa so he can finish the project 1.5 times faster. Terry can therefore finish the project in

$$\frac{2}{1.5} = \frac{4}{3}$$

hours. Now, Melisa can finish $\dfrac{1}{2}$ of the project per hour and Terry can finish $\dfrac{3}{4}$ of the project per hour. Together they can complete

$$\frac{1}{2} + \frac{3}{4} = \frac{5}{4}$$

of the project per hour. Therefore, it would take them $\dfrac{4}{5}$ hours to finish it together.

Problem 5.21 **Janelle cleans her aquarium by replacing $\dfrac{2}{3}$ of the water with new water, but that doesn't clean the aquarium to her satisfaction. She decides to repeat the process, again replacing $\dfrac{2}{3}$ of the water with new water. How many times will Janelle have to do this so that at least 95% of the water is new water?**

Answer

3

Solution

Janelle wants at least 95% of the water to be new water. This is the same as saying she wants less than 5% of the water to be old water. After replacing $\frac{2}{3}$ of the water once, $\frac{1}{3}$ of the water is old water. Replacing the water again,

$$\frac{1}{3} \times \frac{1}{3} = \frac{1}{9} \approx 11.11\%$$

of the water is old water. Replacing the water a third time,

$$\frac{1}{9} \times \frac{1}{3} = \frac{1}{27} \approx 3.70\%$$

of the water is old water. Therefore, she must replace the water three times.

Problem 5.22 **It will take a Type A robot 6 min to weld a fender, but a Type B robot takes only $5\frac{1}{2}$ minutes. If the robots work together for 2 min, how long will it take the Type B robot to finish welding by itself? Express your answer as a mixed number.**

Answer

$1\frac{2}{3}$ minutes

Solution

The Type A robot can weld $\frac{1}{6}$ of a fender in a minute, while the Type B robot can weld

$$\frac{1}{5\frac{1}{2}} = \frac{2}{11}$$

of a fender in a minute. Together, this means that the robots can weld

$$\frac{1}{6} + \frac{2}{11} = \frac{23}{66}$$

of a fender in a minute. Therefore, after two minutes the two robots have finished

$$2 \times \frac{23}{66} = \frac{23}{33}$$

of a fender, so $\dfrac{10}{33}$ remains. It takes the Type B robot

$$\frac{10}{33} \div \frac{2}{11} = \frac{5}{3} = 1\frac{2}{3}$$

minutes to finish the welding itself.

Problem 5.23 **Phil can paint the garage in 12 hours and Rick can do it in 10 hours. They work together for 3 hours. How long will it take Rick to finish the job alone?**

Answer

$\dfrac{9}{2}$ hours

Solution

Phil can paint $\dfrac{1}{12}$ of the garage in 1 hour. Rick can paint $\dfrac{1}{10}$ of the garage in 1 hour. Together they can thus paint

$$\frac{1}{12} + \frac{1}{10} = \frac{11}{60}$$

of the garage in an hour. After 3 hours, they have painted

$$3 \times \frac{11}{60} = \frac{11}{20}$$

of the garage, so $\dfrac{9}{20}$ of it remains. It takes Rick

$$\frac{9}{20} \div \frac{1}{10} = \frac{9}{2}$$

hours to finish painting the garage by himself.

Problem 5.24 **Lincoln can do a job in 8 hours and Dave can do it in 6 hours. What part of the job can they do by working together for 2 hours? For x hours?**

Answer

2 hours: $\dfrac{7}{12}$, x hours: $\dfrac{7x}{24}$

Solution

Lincoln can do $\frac{1}{8}$ of a job in an hour, and Dave can do $\frac{1}{6}$ of a job in an hour. Therefore, they can do

$$\frac{1}{8} + \frac{1}{6} = \frac{7}{24}$$

of a job working together in an hour. Thus, they can do

$$2 \times \frac{7}{24} = \frac{7}{12}$$

of a job in 2 hours, and similarly,

$$x \times \frac{7}{24} = \frac{7x}{24}$$

of a job in x hours.

Problem 5.25 **Andrew spends 5 hours to complete a quarter of a job. Charles spends 6 hours to complete half of the remaining part of the job. How much more time does it take for Andrew and Charles to work together to get the rest of the job done?**

Answer

$\frac{10}{3}$ hours

Solution

Andrew spends 5 hours to complete a quarter of a job, so he can complete

$$\frac{1}{4} \div 5 = \frac{1}{20}$$

of a job in one hour. Hence there is $\frac{3}{4}$ of the job left. Charles spends 6 hours completing half of the remaining part of the job, so Charles completes

$$\frac{3}{4} \times \frac{1}{2} = \frac{3}{8}$$

of the job in 6 hours, so he can complete

$$\frac{3}{8} \div 6 = \frac{1}{16}$$

of a job in one hour. Andrew and Charles can therefore complete

$$\frac{1}{20} + \frac{1}{16} = \frac{9}{80}$$

of a job in one hour. Since there is

$$1 - \frac{1}{4} - \frac{3}{8} = \frac{3}{8}$$

of the job left, it will take Andrew and Charles

$$\frac{3}{8} \div \frac{9}{80} = \frac{10}{3}$$

hours for them to complete the job.

Problem 5.26 **It takes Rianna 24 days to finish a job. For Helen, it takes 32 days. Rianna works on it for some days before Helen takes it over, and it takes a combined total of 26 days to get the job done. How many days does Rianna work on the job?**

Answer

18

Solution 1

(Weighted Average) Rianna can finish $\frac{1}{24}$ of a job in one day, while Helen can finish $\frac{1}{32}$ of a job in one day. The target rate is 1 job in 26 days, or $\frac{1}{26}$ of a job per day. Since this is closer to how much Rianna can work in a day, we know that Rianna will work more days than Helen. We compare the differences of the rates of Rianna and Helen to the target rate. Rianna works

$$\frac{1}{24} - \frac{1}{26} = \frac{1}{312}$$

jobs per day faster than our target, while Helen works

$$\frac{1}{26} - \frac{1}{32} = \frac{3}{416}$$

jobs per day slower. The amount of days Rianna and Helen work will be in the opposite ratio

$$\frac{3}{416} : \frac{1}{312} = 9 : 4.$$

Hence, Rianna works

$$26 \times \frac{9}{13} = 18$$

days (so Helen works the remaining 8).

Solution 2

(Weighted Average Indirect) The target time (26 days) is closer to the time it takes Rianna than the time it takes Helen, so Rianna will complete a higher amount of the job. We compare the differences in time: $26 - 24 = 2$ less days for Rianna and $32 - 26 = 6$ days more for Helen. The amount of the job Rianna and Helen complete will be in the opposite ratio

$$6 : 2 = 3 : 1$$

so Rianna will complete $\frac{3}{4}$ of the job. Since Rianna can complete $\frac{1}{24}$ of the job in a single day, this will take her

$$\frac{3}{4} \div \frac{1}{24} = 18$$

days.

Solution 3

(Algebra) Rianna can finish $\frac{1}{24}$ of a job in one day, while Helen can finish $\frac{1}{32}$ of a job in one day. Let x denote the number of days Rianna works, so $26 - x$ is the number of days Helen works. Since they finish the entire job in these 26 days, we have

$$1 = x \times \frac{1}{24} + (26 - x) \times \frac{1}{32}.$$

Solving the equation,

$$x = 18$$

so Rianna works 18 days and Helen works 8 days.

Problem 5.27 **It takes Jacqueline 50 minutes to type a draft. For Virginia, it takes 30 minutes. Suppose after Jacqueline types for some time, Virginia types the rest of the draft, and it takes a combined total of 42 minutes. What fraction of the draft did Jacqueline type?**

Answer

$\dfrac{3}{5}$

Solution 1

(Weighted Average) The target time, 42 minutes, is closer to the time it takes Jacqueline to type the entire draft, so Jacqueline will complete more of the job. Jacqueline's time is $50 - 42 = 8$ minutes longer than the target, while Virginia's is $42 - 30 = 12$ minutes slower. The amount of the draft Jacqueline and Virginia will be in the opposite ratio

$$12 : 8 = 3 : 2.$$

Therefore, Jacqueline will type $\dfrac{3}{5}$ of the draft.

Solution 2

(Algebra) Let x denote the fraction of the draft Jacqueline types, so $1 - x$ denotes the fraction Virginia types. Therefore, Jacqueline types for $50 \times x$ minutes and Virginia types for $30 \times (1 - x)$ minutes. Since in total it takes 42 minutes to type the draft, we have

$$50x + 30(1 - x) = 42.$$

Solving for x,

$$x = \frac{3}{5},$$

so Jacqueline types $\dfrac{3}{5}$ of the draft.

Problem 5.28 **It takes Elizabeth 9 hours to complete a project. For Tiffany, it takes 12 hours. If they take turns, starting with Elizabeth, each working for one hour at a time, how much total time does it take for them to complete the project?**

Answer

$\dfrac{41}{4}$ hours

Solution

Elizabeth can finish $\dfrac{1}{9}$ of the project in an hour, while Tiffany can finish $\dfrac{1}{12}$. Therefore, in a two hour time period (each working an hour) they can finish

$$\frac{1}{9} + \frac{1}{12} = \frac{7}{36}$$

of the project. Note that $36 \div 7 = 5$ remainder 1, so after 5 such periods (10 hours in total),

$$5 \times \frac{7}{36} = \frac{35}{36}$$

of the project is completed, so $\dfrac{1}{36}$ remains. Since

$$\frac{1}{36} < \frac{1}{9}$$

Elizabeth can finish the rest of the project in less than an hour. Precisely, it takes her

$$\frac{1}{36} \div \frac{1}{9} = \frac{1}{4}$$

of an hour. In total, it takes Elizabeth and Tiffany

$$5 \times 2 + \frac{1}{4} = 10\frac{1}{4} = \frac{41}{4}$$

hours.

Problem 5.29 **If Emily and Julia work together, they can finish a project in 6 days. It takes same amount of time for Emily to complete $\frac{1}{2}$ of the project as it takes for Julia to complete $\frac{1}{3}$ of the project. How long does it take for Emily alone to complete the project?**

Answer

10 days

Solution

We are given in the problem that the ratio of the amount of work each can complete in a given time is

$$\frac{1}{2} : \frac{1}{3} = 3 : 2.$$

Hence, working for the 6 days, Emily completed $\frac{3}{5}$ of the project (and Julia completed $\frac{2}{5}$ of the project). Hence Emily can complete

$$\frac{3}{5} \div 6 = \frac{1}{10}$$

of the project in a day, so it would take her 10 days to finish the project by herself.

Problem 5.30 **Carolyn reads a book. Initially, the number of pages that she had read and the number of pages that she has not read are in ratio $3 : 4$. After she reads an additional 33 pages of the book, the ratio becomes $5 : 3$. How many pages does the whole book have?**

Answer

168

Solution 1

At the start, Carolyn has read $\frac{3}{7}$ of the book. After reading 33 additional pages, she has read $\frac{5}{8}$ of the book. Hence, the 33 pages account for

$$\frac{5}{8} - \frac{3}{7} = \frac{11}{56}$$

of the book. Therefore, the book has

$$33 \div \frac{11}{56} = 168$$

pages in total.

Solution 2

(Algebra) At the start, the number of pages Carolyn has read and not read are in ratio $3 : 4$, so assume she has read $3x$ pages and not read $4x$ pages (so the book has $3x + 4x = 7x$ total pages). After reading 33 more pages, she has read $3x + 33$ and not read $4x - 33$. Since this is in ratio $5 : 3$ we have

$$\frac{3x + 33}{4x - 33} = \frac{5}{3}.$$

Solving for x,

$$x = 24.$$

Therefore the book has $7x = 7 \times 24 = 168$ pages.

Problem 5.31   **Adam can mow his entire yard in three hours. His sister, Brooke, can mow $\frac{3}{4}$ of the same yard in two hours. Using two identical mowers, what part of the yard can they mow in one hour working together?**

Answer

$$\frac{17}{24}$$

Solution

Adam can mow $\frac{1}{3}$ of the yard in one hour. Brooke can mow

$$\frac{3}{4} \div 2 = \frac{3}{8}$$

of the yard in one hour. Together they can therefore mow

$$\frac{1}{3} + \frac{3}{8} = \frac{17}{24}$$

of the yard in one hour.

Problem 5.32 **One pipe can fill a swimming pool 1.5 times faster than a second pipe. If the gardener opens both pipes, they fill the pool in 5 hours. How long would it take to fill the pool if only the slower pipe is used? How about only the faster pipe?**

Answer

$\dfrac{25}{2}$ hours and $\dfrac{25}{3}$ hours

Solution

We are given that the rates that the two pipes fill the pool are in ratio $1.5 : 1 = 3 : 2$. Hence, the amount of water of the pool they fill in 5 hours is in the same ratio $3 : 2$. Therefore, in 5 hours, the first pipe fills $\dfrac{3}{5}$ of the pool, while the second pipe fills $\dfrac{2}{5}$ of the pool. Therefore, the two pipes can respectively fill

$$\frac{3}{5} \div 5 = \frac{3}{25} \text{ and } \frac{2}{5} \div 5 = \frac{2}{25}$$

of the pool in a single hour. Hence it takes the two pipes $\dfrac{25}{3}$ and $\dfrac{25}{2}$ hours to fill the pool separately.

Problem 5.33 **Bill usually takes 50 minutes to groom the horses. After working for 10 minutes, he was joined by Ann and they finished the grooming in 15 minutes. How long would it have taken Ann working alone?**

Answer

20 minutes

Solution

Bill can groom $\dfrac{1}{50}$ of the horses in a single minute. Since Bill works $10 + 15 = 25$ minutes in total, he has groomed

$$25 \times \frac{1}{50} = \frac{1}{2}$$

of the horses. Thus, Ann must have groomed the other $\frac{1}{2}$ of the horses in 10 minutes, so it would take her $10 \times 2 = 20$ minutes to groom all the horses working alone.

Problem 5.34 **Mona can complete a task alone in 150 minutes. Sarah can finish the same task in 3 hours. They work together for 30 minutes, and then a new worker, Li, joins them, and they finish the task 30 minutes later. How long would it take Li to finish the task alone?**

Answer

$\frac{225}{2}$ minutes

Solution

We know that working alone Mona can complete $\frac{1}{150}$ of the task in one minute. Since 3 hours is 180 minutes, Sarah can finish $\frac{1}{180}$ of the task in one minute. In the question above, Mona and Sarah each work a total of $30 + 30 = 60$ minutes. Therefore, Mona completes

$$60 \times \frac{1}{150} = \frac{2}{5}$$

of the task and Sarah completes

$$60 \times \frac{1}{180} = \frac{1}{3}$$

of the task. Therefore, Li, who works a total of 30 minutes, completes

$$1 - \frac{2}{5} - \frac{1}{3} = \frac{4}{15}$$

of the task, so she can complete

$$\frac{4}{15} \div 30 = \frac{2}{225}$$

of the task in one minute. Hence Li would take $\frac{225}{2}$ minutes to complete the task herself.

Problem 5.35 **If pump A is used alone, it takes 6 hours to fill the pool. Pump B takes 8 hours alone to fill the same pool. Uncle Sam wants to use three pumps: A, B and C to fill the pool in 2 hours. What should be the rate of pump C in order to accomplish Uncle Sam's goal?**

Answer

$\dfrac{5}{24}$ of the pool per hour

Solution 1

The rate of pump A is $\dfrac{1}{6}$ of the pool per hour and the rate of pump B is $\dfrac{1}{8}$ of the pool per hour. Therefore, in two hours, pump A fills

$$2 \times \frac{1}{6} = \frac{1}{3}$$

of the pool and pump B fills

$$2 \times \frac{1}{8} = \frac{1}{4}$$

of the pool. Hence pump C must fill

$$1 - \frac{1}{3} - \frac{1}{4} = \frac{5}{12}$$

of the pool in two hours. Hence pump C must work at a rate of

$$\frac{5}{12} \div 2 = \frac{5}{24}$$

of the pool per hour.

Solution 2

(Algebra) The rate of pump A is $\dfrac{1}{6}$ of the pool per hour and the rate of pump B is $\dfrac{1}{8}$ of the pool per hour. If x is the rate of pump C, the the three pumps working together fill

$$\frac{1}{6} + \frac{1}{8} + x = \frac{7}{24} + x$$

of the pool per hour. Therefore, if the pumps fill the entire pool in two hours, we have

$$1 = 2 \times \left(\frac{7}{24} + x \right) = \frac{7}{12} + 2x.$$

Solving for x,

$$x = \frac{5}{24},$$

where x is the rate of pump C as a fraction of the pool per hour.

Problem 5.36 **Patty and Tracy can finish decorating a house for the holidays in 2.5 hours if they work together. Patty works twice as fast as Tracy. How long would it take to each of them if they work alone?**

Answer

Patty: $\dfrac{15}{4}$ hours, Tracy: $\dfrac{15}{2}$ hours

Solution

Since Patty works twice as fast as Tracy (in a ratio of $2 : 1$) the amount of the house they have each decorated at the end of 2.5 hours is in ratio $2 : 1$ as well. Therefore, Patty has decorated $\dfrac{2}{3}$ of the house while Tracy has decorated $\dfrac{1}{3}$ in

$$2.5 = \frac{5}{2}$$

hours. Hence, Patty can decorate

$$\frac{2}{3} \div \frac{5}{2} = \frac{4}{15}$$

of the house per hour, while Tracy can decorate

$$\frac{1}{3} \div \frac{5}{2} = \frac{2}{15}$$

of the house per hour. Thus, it takes Patty $\dfrac{15}{4}$ hours to decorate the entire house and Tracy $\frac{15}{2}$ hours.

Problem 5.37   **Peter can paint a wall in 40 minutes and John can paint the wall in 60 minutes. If they work together for 12 minutes, how much of the wall is left unpainted?**

Answer

$$\frac{1}{2}$$

Solution

Peter can paint the wall in 40 minutes, so he can paint $\frac{1}{40}$ of a wall in one minute. Similarly, John can paint $\frac{1}{60}$ of the wall in one minute. Working together they can paint

$$\frac{1}{40} + \frac{1}{60} = \frac{1}{24}$$

of the wall in one minute. Working together for 12 minutes they can paint

$$12 \times \frac{1}{24} = \frac{1}{2}$$

of the wall. Therefore, $\frac{1}{2}$ of the job remained after 12 minutes of working together.

Problem 5.38   **James, Patty, and Joseph can organize the new products in the warehouse in 2 hours. If James does the job alone, he can finish in 5 hours. If Patty does the job alone, she can finish it in 6 hours. How long will it take for Joseph to finish the job alone?**

Answer

$$\frac{15}{2} \text{ hours}$$

Solution

James can finish $\frac{1}{5}$ of the job himself in an hour, and Patty can finish

$\dfrac{1}{6}$ of the job herself in an hour. Hence, working together for 2 hours they can complete

$$2 \times \left(\dfrac{1}{5} + \dfrac{1}{6}\right) = \dfrac{11}{15}$$

of the job. In the two hours all three worked together, Joseph must have completed the remaining $\dfrac{4}{15}$ of the job. Joseph can therefore complete

$$\dfrac{4}{15} \div 2 = \dfrac{2}{15}$$

of the organizing himself on one hour. Thus, it would take him $\dfrac{15}{2}$ hours to complete the entire job alone.

Problem 5.39 **There is a drain and a hose in the pool. It is known that the hose can fill the pool in 21 hours, and the drain can empty the pool in 24 hours. How long does it take to fill the pool if the drain is open at the same time?**

Answer

168 hours

Solution 1

The hose fills the pool at a rate of $\dfrac{1}{21}$ of the pool per hour while the drain empties the pool at a rate of $\dfrac{1}{24}$ of the pool per hour. Since the hose is faster than the drain, if the hose is on and the drain is open, the pool is being filled at a rate of
$$\dfrac{1}{21} - \dfrac{1}{24} = \dfrac{1}{168}$$
of the pool per hour. It will then take 168 hours to fill the pool.

Solution 2

The hose fills the pool 3 hours faster than the drain empties it. Therefore, after filling a pool
$$21 \div 3 = 7$$

times, the hose can fill one more pool than the drain can empty in the same time. Hence, after $7 \times 24 = 168$ hours, the pool will be full if the hose and drain are open at the same time.

Problem 5.40 A senior worker and a new worker together produce a set of machines. The senior worker can produce 40 pieces per hour, and the new worker can produce 30 pieces per hour. When they finished, the new worker produced exactly 450 pieces. How many pieces did they produce in total?

Answer

1050

Solution 1

It takes the new worker

$$450 \div 30 = 15$$

hours to produce 450 pieces. In those 15 hours, the senior worker produced

$$15 \times 40 = 600$$

pieces. In total $600 + 450 = 1050$ pieces were produced.

Solution 2

Since the senior worker's and new worker's rate of production is in ratio

$$40 : 30 = 4 : 3,$$

the amount of pieces they produce is in the same ratio. Multiplying both sides of this ratio by

$$450 \div 3 = 150$$

we have

$$4 : 3 = 600 : 450$$

so the senior worker produces 600 pieces if the new worker produces 450. Together the two workers produce $600 + 450 = 1050$ pieces in total.

Problem 5.41 **When two teams A and B work together, it takes 18 days to get a job completed. After team A works for 3 days, and team B works for 4 days, only $\frac{1}{5}$ of the job is done. How long does it take for team A alone to complete the job? For team B alone?**

Answer

A: 45 days, B: 30 days

Solution 1

We know that if team A works for 3 days and team B works for 4 days they can finish $\frac{1}{5}$ of the job. If they work 6 times longer, team A works for 18 days and team B works for 24 days and they complete $\frac{6}{5}$ of a job. Combined with the fact that the two teams can complete the job in 18 days, we have that if team B works $24 - 18 = 6$ days it can complete $\frac{6}{5} - 1 = \frac{1}{5}$ of a job. Thus team B can complete

$$\frac{1}{5} \div 6 = \frac{1}{30}$$

of a job in one day. In 18 days team B completes

$$18 \times \frac{1}{30} = \frac{3}{5}$$

of a job, so team A must complete $\frac{2}{5}$ of a job in 18 days, and

$$\frac{2}{5} \div 18 = \frac{1}{45}$$

of a job in one day. We then know it takes team A 45 days and team B 30 days to complete the job if they work alone.

Solution 2

(Algebra) Let x denote the amount of work team A can complete in one day and y the amount of work team B can complete in one day. Working together they can complete the job in 18 days so

$$18 \times x + 18 \times y = 1.$$

If team A works for 3 and team B works for 4 days, $\dfrac{1}{5}$ of the job is completed, so

$$3 \times x + 4 \times y = \frac{1}{5}.$$

This gives us the system of equations

$$\begin{cases} 18x + 18y & = & 1, \\ 3x + 4y & = & \frac{1}{5}. \end{cases}$$

Multiplying the second equation by 6, we get

$$18x + 24y = \frac{6}{5}.$$

Subtracting the first equation from this equation we have

$$6y = \frac{1}{5}.$$

so

$$y = \frac{1}{30}.$$

Substituting back into the first equation,

$$18x + 18 \times \frac{1}{30} = 1,$$

so we can solve for x to get

$$x = \frac{1}{45}.$$

Therefore it takes team A 45 days working alone and team B 30 days working alone to complete the job.

Problem 5.42 **Suppose if Ethan works for 5 days and Owen for 6 days, they finish a project. Alternatively, if Ethan works for 7 days and Owen for 2 days, they also finish the project. How long does it take for Ethan alone to complete the project? For Owen alone?**

Answer

Ethan: 8 days, Owen: 16 days

Solution 1

Since Ethan working 7 days and Owen working 2 days is enough to complete the project, if Ethan works $7 \times 3 = 21$ days and Owen works $2 \times 3 = 6$ days, they could complete 3 projects. We also know that if Ethan works 5 days and Owen works 6 days they can complete one project. Since Owen works for 6 days in both cases, in $21 - 5 = 16$ days working alone, Ethan can finish 2 projects. Therefore, it takes Ethan 8 days to finish one project himself. Thus, in 7 days Ethan can finish

$$7 \div 8 = \frac{7}{8}$$

of the project. Hence, Owen can finish $\frac{1}{8}$ of the project in 2 days. Thus it takes Owen

$$2 \div \frac{1}{8} = 16$$

to finish a project himself.

Solution 2

Let x be the work rate of Ethan and y the work rate for Owen. If Ethan works for 5 days and Owen for 6 days, they finish the project so

$$5 \times x + 6 \times y = 1.$$

They also finish the project if Ethan works for 7 days and Owen works 2 days so

$$7 \times x + 2 \times y = 1$$

This gives us the system of equations

$$\begin{cases} 5x + 6y &= 1, \\ 7x + 2y &= 1. \end{cases}$$

Multiplying the second equation by 3, we have

$$21x + 6y = 3.$$

Subtracting the first equation from this we have

$$16x = 2,$$

so

$$x = \frac{1}{8}.$$

Substituting back into the first equation,

$$5 \times \frac{1}{8} + 6y = 1,$$

so we can solve for y, and

$$y = \frac{1}{16}.$$

Therefore, Ethan can complete the project in 8 days himself, while it takes Owen 16 days to complete the project working alone.

Problem 5.43 **Students from the Key Club at Whitman High School wash cars from the two parking lots A and B. There are four times as many cars in lot A than in lot B. First, they wash the cars in lot A for half a day. Next, half of students continue, and half of students start to wash cars in lot B. The work in lot B is done at the end of the day. Unfortunately, there are still cars unwashed in lot A. If all the students worked together to finish the cars in lot A, how long would it take?**

Answer

$\frac{1}{4}$ of a day

Solution

(Algebra) Let x be the fraction of the total cars that the whole collection of students can wash in a full day. Since the number of cars in the two lots is in ratio $4 : 1$, $\frac{4}{5}$ of the cars are in lot A and $\frac{1}{5}$ of the cars are in lot B. Since $\frac{1}{2}$ of the students can wash the cars in lot B (which is $\frac{1}{5}$ of the total cars) in $\frac{1}{2}$ of the day, we have

$$\frac{x}{2} \times \frac{1}{2} = \frac{1}{5}.$$

Solving for x,

$$x = \frac{4}{5},$$

so all the students working together can walk $\frac{4}{5}$ of the total cards in both lots. Since all the students worked together in the morning in lot A and half of them continued in the afternoon the amount of cars washed in lot A in total was

$$x \times \frac{1}{2} + \frac{x}{2} \times \frac{1}{2} = \frac{3}{4} \times x = \frac{3}{4} \times \frac{4}{5} = \frac{3}{5}$$

of the total cars. Since lot A has $\frac{4}{5}$ of the total cars, $\frac{1}{5}$ of them remain. It will therefore take the students

$$\frac{1}{5} \div x = \frac{1}{5} \div \frac{4}{5} = \frac{1}{4}$$

of an extra day to finish the cars in lot A together.

Problem 5.44 **If Iris spends 3 days and Olivia spends 5 days on a project, $\frac{1}{2}$ of the work can be completed. If instead Iris spends 5 days and Olivia spends 3 days on the project, $\frac{1}{3}$ of the work can be completed. How long does it take to complete the whole project if Iris and Olivia work together?**

Answer

Iris: 96 days, Olivia: $\frac{32}{3}$ days

Since Iris and Olivia can finish half the project working for 3 and 5 days each and they can complete a third of the project working for 5 and 3 days each, if they each work a total of $3 + 5 = 8$ days, they can complete

$$\frac{1}{2} + \frac{1}{3} = \frac{5}{6}$$

of the project. Therefore they can complete

$$\frac{5}{6} \div 8 = \frac{5}{48}$$

of the project in a single day so it takes them $\dfrac{48}{5}$ days to finish the project working together.

Solution 2

(Algebra) Let x denote the fraction of the project Iris can complete in a day and y denote the fraction Olivia can complete in a day. If Iris works 3 days and Olivia 5, they finish half the project, so

$$3 \times x + 5 \times y = \frac{1}{2}.$$

Similarly, if Iris works 5 and Olivia 3 days, they finish one-third of the project, so

$$5 \times x + 3 \times y = \frac{1}{3}.$$

This gives the system of equations

$$\begin{cases} 3x + 5y &=& \frac{1}{2}, \\ 5x + 3y &=& \frac{1}{3}. \end{cases}$$

We can use this system to solve for x and y separately, but for the question we really want to know the value of $x+y$. Adding the two equations together we have

$$8x + 8y = \frac{5}{6}$$

so dividing by 8 we have

$$x + y = \frac{5}{48}$$

where $x+y$ is the amount of the project Iris and Olivia can complete working together in a single day. Therefore, it takes them $\dfrac{48}{5}$ days to complete the whole project together.

Problem 5.45 **A pool has an inlet pump and an outlet pump. If the pool is empty, and the inlet pump is open, it takes 5 hours to fill the pool with water. If the pool is full, and the outlet pump is opened, it takes 7 hours to empty the pool. Suppose after the inlet pump is open for 2 hours, both the pumps are opened. How much longer does it take for the pool to be half full of water?**

Answer

$\dfrac{7}{4}$ hours

Solution

The inlet pump can fill $\dfrac{1}{5}$ of the pool in one hour, while the outlet pump can drain $\dfrac{1}{7}$ of the pool in an hour. Since the inlet pump is faster, the pool is filled

$$\frac{1}{5} - \frac{1}{7} = \frac{2}{35}$$

in one hour if both pumps are open. If the inlet pump is open for two hours,

$$2 \times \frac{1}{5} = \frac{2}{5}$$

of the pool is filled, so

$$\frac{1}{2} - \frac{2}{5} = \frac{1}{10}$$

remains if the pool needs to be half full. With both pumps open, it will take

$$\frac{1}{10} \div \frac{2}{35} = \frac{7}{4}$$

more hours to fill the pool halfway.

Problem 5.46 It takes $\frac{1}{3}$ more time for Andy to plant one tree than for Nathan. If Andy and Nathan work together, then in the end Nathan plants 36 more trees than Andy does. How many trees are in total?

Answer

252

Solution 1

Andy takes $\dfrac{1}{3}$ more time to plant a tree. Since no time is mentioned at all in the problem, we may assume it takes Nathan 3 minutes to plant a tree, so it takes 4 minutes for Andy to plant a tree. Therefore, in 12 minutes, Nathan can plant 1 more tree than Andy (as Andy can plant 3 trees and Nathan can plant 4 trees). Therefore, in

$$12 \times 36 = 432$$

minutes, Nathan plants 36 more trees than Andy. In this time Nathan plants

$$4 \times 36 = 144$$

trees and Andy plants

$$3 \times 36 = 108$$

trees. A total of $144 + 108 = 252$ trees are planted.

Solution 2

(Algebra) Since it takes Andy $\frac{1}{3}$ more time to plant a tree than Nathan, the ratio of the time it takes them to plant trees is $4 : 3$. Therefore, the amount of trees they plant is in ratio $3 : 4$. Let x be such that Andy plants $3x$ trees and Nathan plants $4x$ trees. Since Nathan plants 36 more trees we have

$$3x + 36 = 4x.$$

Solving for x,

$$x = 36.$$

Therefore, there are $7x = 7 \times 36 = 252$ trees planted in total.

Problem 5.47 Anthony can cut a lawn in 2 hours, Mia can cut the same lawn in 3 hours, and Dandria can cut the same lawn in 2 hours. Anthony cuts the lawn for $\frac{1}{2}$ hour, and then Mia replaces Anthony and cuts the lawn for 1 hour herself. How many additional minutes will it take Dandria to finish cutting the lawn by herself?

Answer

50

Solution

Anthony can cut $\frac{1}{2}$ of the lawn in one hour, Mia can cut $\frac{1}{3}$ of the lawn in one hour, and Dandria can also cut $\frac{1}{2}$ of the lawn in one hour. Therefore, if Anthony cuts the lawn for $\frac{1}{2}$ hour, he cuts

$$\frac{1}{2} \times \frac{1}{2} = \frac{1}{4}$$

of the lawn. Since Mia can cut $\frac{1}{3}$ of the lawn in an hour, Anthony and Mia cut

$$\frac{1}{4} + \frac{1}{3} = \frac{7}{12}$$

of the lawn. Dandria will then need to mow $\frac{5}{12}$ of the lawn, which takes her

$$\frac{5}{12} \div \frac{1}{2} = \frac{5}{6}$$

of an hour, or 50 minutes, to finish.

Problem 5.48 A pool has two inlet pumps A and B. If pump A alone is open, it takes 12 hours to fill the pool with water. If pump B is open, it takes 18 hours to fill the pool with water. If the pool needs to be filled in 10 hours, what is the least amount of time both pumps need to be open?

Answer

3 hours

Solution 1

Pump A fills $\frac{1}{12}$ of the pool in one hour, while pump B fills $\frac{1}{18}$ of the pool in one hour. Therefore, the pumps working together fill

$$\frac{1}{12} + \frac{1}{18} = \frac{5}{36}$$

of the pool in one hour. Hence, the two pumps could fill

$$\frac{5}{36} \times 10 = \frac{50}{36}$$

of the pool in 10 hours. Since this overfills the pool by $\frac{7}{18}$, we can turn of one of the pumps for some time. If we want to have both pumps on together for the smallest possible amount of time, we turn off the slower pump (pump B). It takes pump B

$$\frac{7}{18} \div \frac{1}{18} = 7$$

hours to account for the overfilled water, so we can turn pump B off for up to 7 hours. Therefore, the two pumps must work together for at least $10 - 7 = 3$ hours.

Solution 2

(Algebra) Pump A fills $\frac{1}{12}$ of the pool in one hour, while pump B fills $\frac{1}{18}$ of the pool in one hour. Since pump A is faster, the least amount of time the two pumps work together will occur when either both pumps or just pump A are working at all times. Let x denote the amount of time both pumps work together, so $10 - x$ is the amount of time only pump A is working. Since the two pumps can fill

$$\frac{1}{12} + \frac{1}{18} = \frac{5}{36}$$

of the pool in one hour, if the pool will be filled in 10 hours we have

$$\frac{5}{36} \times x + \frac{1}{12} \times (10 - x) = 1.$$

Solving for x,

$$x = 3$$

so the pumps must work together at least 3 hours.

Problem 5.49 **There are two pumps A and B. They are used to fill water in two pools that, if full, hold equal amount of water. The ratio of the water-pumping**

rate for pump A and pump B is $7 : 5$. **After $2\frac{1}{3}$ hours, the water in the two pools, if combined, can exactly fill one pool. Next pump A increases the water-pumping rate by 25%, pump B reduces the water-pumping rate by 30%. After pump A fills the pool, how much longer does it take for pump B to fill the second pool? Express your answer as a mixed number.**

Answer

$3\dfrac{1}{3}$ hours

Solution

Since the original ratio of the water-pumping rates is $7 : 5$, the amount of water each pump fills in a given amount of time is in ratio $7 : 5$. Therefore, pump A fills $\dfrac{7}{12}$ and pump B fills $\dfrac{5}{12}$ of a pool in the initial $2\dfrac{1}{3} = \dfrac{7}{3}$ hours. We can then calculate that initially pump A fills

$$\frac{7}{12} \div \frac{7}{3} = \frac{1}{4}$$

of one pool per hour and pump B fills

$$\frac{5}{12} \div \frac{7}{3} = \frac{5}{28}$$

of one pool per hour. Since pump A's rate then increases to

$$\frac{1}{4} \times (1 + 25\%) = \frac{1}{4} \times \frac{5}{4} = \frac{5}{16}$$

of a pool per hour, it takes pump A an additional

$$\frac{5}{12} \div \frac{5}{16} = \frac{4}{3}$$

hours to fill the first pool. Pump B's rate decreases to

$$\frac{5}{28} \times (1 - 30\%) = \frac{5}{28} \times \frac{7}{10} = \frac{1}{8}$$

of a pool per hour. Thus, it takes pump B an additional

$$\frac{7}{12} \div \frac{1}{8} = \frac{14}{3}$$

hours to fill the second pool. Hence pump B takes

$$\frac{14}{3} - \frac{4}{3} = \frac{10}{3}$$

longer to fill the second pool.

Problem 5.50 **Brandon, Richard and Samuel are friends. They build a wall divider in a yard. Brandon and Richard work together, and they build $\frac{1}{3}$ of the divider in 5 days. Next, Richard and Samuel work together, and they build $\frac{1}{4}$ of the rest of the divider in 2 days. Last, Brandon and Samuel work together, they finish the rest in 5 days. How long does it take for Brandon alone to build the divider? For Richard alone? For Samuel alone?**

Answer

Brandon: 24 days, Richard: 40 days, Samuel: $\dfrac{120}{7}$ days.

Solution 1

Brandon and Richard can build $\frac{1}{3}$ of the wall divider in 5 days, so they can build

$$\frac{1}{3} \div 5 = \frac{1}{15}$$

of the divider in one day. There is $\frac{2}{3}$ of the divider left, so in 2 days Richard and Samuel build

$$\frac{1}{4} \times \frac{2}{3} = \frac{1}{6}$$

of the divider. In one day Richard and Samuel can therefore build

$$\frac{1}{6} \div 2 = \frac{1}{12}$$

of the dividier. As

$$\frac{1}{3} + \frac{1}{6} = \frac{1}{2}$$

Brandon and Samuel build $\frac{1}{2}$ of the divider in 5 days, so they can build

$$\frac{1}{2} \div 5 = \frac{1}{10}$$

---

of the divider in one day. Note these 3 pieces of information tell us that if Brandon, Richard, and Samuel each work for 2 days, all together they can build

$$\frac{1}{15} + \frac{1}{12} + \frac{1}{10} = \frac{1}{4}$$

of the divider. Therefore, in one day working all together they can build

$$\frac{1}{4} \div 2 = \frac{1}{8}$$

of the divider. Recalling that Brandon and Richard can build $\frac{1}{15}$ of the divider in one day, we have the Samuel himself can build

$$\frac{1}{8} - \frac{1}{15} = \frac{7}{120}$$

of the divider in one day. Similarly, Richard can build

$$\frac{1}{8} - \frac{1}{10} = \frac{1}{40}$$

of the divider in one day. Lastly, Brandon can build

$$\frac{1}{8} - \frac{1}{12} = \frac{1}{24}$$

of the divider in one day. Therefore, it takes Brandon 24 days, Richard 40 days, and Samuel $\frac{120}{7} \approx 17.14$ days to build the divider alone.

Solution 2

Let x be how much of the divider Brandon can build in one day, y be how much Richard can build in one day, and z be how much Samuel can build in one day. Since Brandon and Richard can build $\frac{1}{3}$ of the divider in 5 days, we have

$$5x + 5y = \frac{1}{3}.$$

There is $\frac{2}{3}$ of the divider left, so in 2 days Richard and Samuel build

$$\frac{1}{4} \times \frac{2}{3} = \frac{1}{6}$$

of the divider and therefore

$$2y + 2z = \frac{1}{6}.$$

Since

$$\frac{1}{3} + \frac{1}{6} = \frac{1}{2}$$

Brandon and Samuel build the second half of the divider in 5 days, so

$$5x + 5z = \frac{1}{2}.$$

We have the system of equations

$$\begin{cases} 5x + 5y &= \dfrac{1}{3}, \\ 2y + 2z &= \dfrac{1}{6}, \\ 5x + 5z &= \dfrac{1}{2}. \end{cases}$$

Dividing the equations by 5, 2, and 5, we get a new system

$$\begin{cases} x + y &= \dfrac{1}{15}, \\ y + z &= \dfrac{1}{12}, \\ x + z &= \dfrac{1}{10}. \end{cases}$$

Adding all three up and dividing by 2, we get

$$x + y + z = \left(\frac{1}{15} + \frac{1}{12} + \frac{1}{10} \right) \div 2 = \frac{1}{8}.$$

We can subtract the 3 equations the second system of equations above from this one to solve for z, x, and y,

$$x = \frac{1}{8} - \frac{1}{12} = \frac{1}{24}, \qquad y = \frac{1}{8} - \frac{1}{10} = \frac{1}{40}, \qquad z = \frac{1}{8} - \frac{1}{15} = \frac{7}{120}.$$

Therefore it takes Brandon 24 days to build the divider himself, Richard 40 days to build the divider himself, and Samuel $\dfrac{120}{7} \approx 17.14$ days to build the divider himself.

Problem 5.51 **A certain number of small parts need to be produced. 30 parts are scheduled to be produced after each day. After $\dfrac{1}{3}$ of the parts are produced, the rate of production increases by 10% thanks to improvement in efficiency. It takes 4 fewer days to produce all the parts than scheduled. How many parts are in total?**

Answer

1980

Solution 1

Originally 30 parts are scheduled to be produced each day. A 10% increase in efficiency means that

$$30 \times 10\% = 30 \times 0.1 = 3$$

extra parts are produced each day. For the project to finish 4 days early, we need a total of

$$4 \times 30 = 120$$

extra parts. With the increased efficiency, this takes

$$120 \div 3 = 40$$

days. In these 40 days, a total of

$$40 \times 33 = 1320$$

parts are made. However, recall only $\dfrac{2}{3}$ of the parts were made with increased efficiency. Therefore,

$$1320 \div \frac{2}{3} = 1980$$

parts are produced in total.

Solution 2

(Algebra) Let x be the number of parts in total. Since originally 30 parts are scheduled to be produced each day, $\dfrac{x}{30}$ days are scheduled to produce the parts. It takes

$$\frac{x}{3} \div 30 = \frac{x}{90}$$

days to produce the first $\dfrac{1}{3}$ of the parts. With a 10% improvement in efficiency,

$$30 \times (1 + 10\%) = 30 \times 1.1 = 33$$

parts can be produced each day. It thus takes

$$\frac{2x}{3} \div 33 = \frac{2x}{99}$$

days to produce the remaining $\dfrac{2}{3}$ of the parts. Since this takes 4 days less than originally scheduled, we have the equation

$$\frac{x}{90} + \frac{2x}{99} = \frac{x}{30} - 4.$$

Solving for x,

$$x = 1980$$

so there are 1980 parts in total.

Problem 5.52   **Peter painted $\dfrac{1}{3}$ of a room while Richard painted $\dfrac{2}{5}$ of the same room. It then took Peter 1 hour, 40 minutes to finish painting the remainder of the room by himself. In how many hours could Peter have painted the entire room by himself? Express your answer as a mixed number.**

Answer

$6\dfrac{1}{4}$.

Solution

Peter paints $\dfrac{1}{3}$ of the room and Richard paints $\dfrac{2}{5}$ of the room, so

$$\frac{1}{3} + \frac{2}{5} = \frac{11}{15}$$

of the room is painted and $\frac{4}{15}$ left to paint. Since it takes Peter 1 hour and 40 minutes, or $1\frac{2}{3} = \frac{5}{3}$ hours to finish painting the room, Peter can paint

$$\frac{4}{15} \div \frac{5}{3} = \frac{4}{25}$$

of the room in one hour. Therefore it takes Peter $\frac{25}{4} = 6\frac{1}{4}$ hours to paint the entire room.

Problem 5.53 **Adam and Bob are each assigned a task to paint a wall. The two walls are identical. At the beginning, Adam went to the wrong wall and painted 500 square feet on Bob's wall. At this moment Bob came and found Adam's mistake. Adam then returned to his own wall and Bob continued painting his wall. After a few days, Bob finished his task, but Adam is not yet done with his job. Now Bob decided to come and help Adam. Bob painted 1000 square feet on Adam's wall. Which person did more of the job?**

Answer

Bob

Solution

(Algebra) Suppose each wall is x square feet in total. Looking at Bob's wall first, Adam paints 500 square feet, and Bob paints the rest, so Bob paints $x - 500$ square feet. For Adam's wall, Bob paints 1000 square feet, so Adam paints the remaining $x - 1000$ feet. Therefore, in total Adam paints

$$500 + x - 1000 = x - 500$$

square feet, while Bob paints

$$x - 500 + 1000 = x + 500$$

square feet. Therefore, Bob has painted more and done more of the job.

Problem 5.54 **A pool can be filled by pipe A in 3 hours, and pipe B in 5 hours. When the pool is full, it can be drained by pipe C in 4 hours. Suppose the pool is empty and all three pipes are open, how long will it take to fill up the pool?**

Answer

$$\frac{60}{17}.$$

Solution

Since pipe A fills the pool in 3 hours, pipe A can fill $\frac{1}{3}$ of the pool in one hour. Similarly, pipe B fills the pool in 5 hours, so fills $\frac{1}{5}$ of the pool in one hour. Lastly, pipe C drains the pool in 4 hours, so drains $\frac{1}{4}$ of the pool in one hour. Therefore, if all three pipes are open, they fill

$$\frac{1}{3}+\frac{1}{5}-\frac{1}{4}=\frac{17}{60}$$

of the pool in one hour. Therefore, the pool is filled in $\frac{60}{17}\approx 3.53$ hours.

Problem 5.55 **Robot A can produce 48 pieces of machinery per hour, and robot B can produce 36 pieces per hour. After they worked together for 8 hours, there were 64 pieces marked as defective when the company tested them. How many non-defective pieces they can produce together in one hour?**

Answer

76

Solution

Working together, robots A and B produce $48+36=84$ pieces per hour. However, since 64 pieces are marked defective in 8 hours,

$$64\div 8=8$$

pieces are marked defective in 1 hour. Therefore, together they produce $84-8=76$ non-defective pieces per hour.

Problem 5.56 **Company A and B plan to manufacture a number of TVs together. After Company A worked for 6 days, it had finished $\frac{1}{4}$ of the TVs. Then the two**

companies worked together and finished the rest of the TVs in 6 days. It is known that Company B can produce 80 TVs per day. How many TVs in total did the two companies produce?

Answer

960

Solution

In total, Company A works for $6 + 6 = 12$ days and Company B works for 6 days to complete all the TVs. Since Company A completes $\frac{1}{4}$ of the TVs in 6 days, it completes

$$\frac{1}{4} \times 2 = \frac{1}{2}$$

of the TVs in 12 days. Therefore Company B completes $\frac{1}{2}$ of the TVs in 6 days. Since Company B can produce 80 TVs per day, they complete

$$80 \times 6 = 480$$

TVs in 6 days. In total the two companies produced $480 \times 2 = 960$ TVs.

Problem 5.57   Ben and Jack can finish a task in 6 days working together. If Ben works alone for 5 days and then Jack takes over and works for 3 days, they finish $\frac{7}{10}$ of the work. How long would it take for each of them complete the task working alone?

Answer

Ben: 10 days, Jack: 15 days

Solution 1

Ben and Jack can finish the task each working a total of 6 days. They can also finish $\frac{7}{10}$ of a task if Ben works 5 days and Jack works 3 days, so they can finish

$$\frac{7}{10} \times 2 = \frac{7}{5}$$

of a task if Ben works 10 days and Jack works 6 days. Since Jack works a total of 6 days in both of the situations above, we see Ben can finish

$$\frac{7}{5} - 1 = \frac{2}{5}$$

of the task in $10 - 6 = 4$ days. Therefore, Ben can finish

$$\frac{2}{5} \div 4 = \frac{1}{10}$$

of the task in one day. In 6 days he can finish

$$\frac{1}{10} \times 6 = \frac{3}{5}$$

of the job, so Jack can finish $\frac{2}{5}$ of the job in 6 days. Jack can thus finish

$$\frac{2}{6} \div 6 = \frac{1}{15}$$

of the job in one day. Hence Ben can finish a task in 10 days working himself, while it takes Jack 15 days working by himself to finish the task.

Solution 2

Let x be the amount of a task Ben can finish in one day and y the amount Jack can finish in one day (each working by themselves). Since they take 6 days to finish a task working together

$$6 \times x + 6 \times y = 1.$$

Since they can finish $\frac{7}{10}$ of a task if Ben works 5 days and Jack works 3, we have

$$5 \times x + 3 \times y = \frac{7}{10}.$$

This gives the system of equations

$$\begin{cases} 6x + 6y &= 1, \\ 5x + 3y &= \frac{7}{10}. \end{cases}$$

Multiplying the second equation by 2 gives

$$10x + 6y = \frac{7}{5}.$$

Subtracting the first equation from this we have

$$4x = \frac{2}{5}$$

so we can solve for x,

$$x = \frac{1}{10}.$$

Substituing into the first equation we have

$$6 \times \frac{1}{10} + 6y = 1$$

so we can solve for y to get

$$y = \frac{1}{15}.$$

Therefore, Ben can finish a task working 10 days by himself, while it takes Jack 15 days working by himself.

Problem 5.58 **A senior worker and a new worker worked together for 2 days and finished $\frac{3}{5}$ of their work. The senior worker then took 2 days off while the new worker continued. After that, the senior worker went back the two workers worked together to finish their work. If the senior worker works twice as fast as the new worker, how many days did it take in total to finish the work? Express your answer as a mixed number.**

Answer

$4\frac{2}{3}.$

Solution

The senior worker works twice as fast as the new worker, so he does twice as much of the work in the same amount of time. Therefore, as

$$\frac{2}{5} + \frac{1}{5} = \frac{3}{5}$$

in the first two days the senior worker completed $\frac{2}{5}$ of the work and the new worker completed $\frac{1}{5}$ of the work. This means during the two days the senior worker took off, the new worker completes another $\frac{1}{5}$ of the job. There is thus $\frac{1}{5}$ of the job remaining. Since working together the two workers can complete $\frac{3}{5}$ of the job in two days, they can complete

$$\frac{3}{5} \div 2 = \frac{3}{10}$$

of the job in one day. It then takes

$$\frac{1}{5} \div \frac{3}{10} = \frac{2}{3}$$

of a day to complete the last $\frac{1}{5}$ of the job. In total they finish the work in $2 + 2 + \frac{2}{3} = 4\frac{2}{3}$ days.

Problem 5.59 **Calvin and Tony worked on producing a set of machines. Calvin planned to complete $\frac{7}{12}$ of the task. After he finished, he helped Tony to produce 24 pieces. The ratio of the number of pieces of Calvin to Tony is 5:3. How many pieces did Tony make?**

Answer

216

Solution 1

(Algebra) Suppose there are $8x$ pieces in total, so Calvin makes $5x$ pieces and Tony makes $3x$ pieces. We also know that Calvin makes $\frac{7}{12}$ of the total pieces plus 24 extra, giving the equation so we also know that Calvin makes

$$\frac{7}{12} \times 8x + 24 = \frac{14x}{3} + 24$$

pieces in total. This gives us the equation

$$\frac{14x}{3} + 24 = 5x$$

so solving for x we get

$$x = 72.$$

Therefore, Tony makes $3x = 3 \times 72 = 216$ pieces in total.

Solution 2

Calvin originally planned to make $\dfrac{7}{12}$ of the total pieces. However, he ended up making $\frac{5}{8}$ of the total pieces. Therefore he made an extra

$$\frac{5}{8} - \frac{7}{12} = \frac{1}{24}$$

of the total pieces. Since we know this extra amounts to 24 pieces, there are

$$24 \div \frac{1}{24} = 576$$

pieces made in total. Since Calvin makes $\dfrac{3}{8}$ of the total pieces, he makes

$$576 \times \frac{3}{8} = 216$$

pieces in total.

Problem 5.60 **A project is planned to be completed by 45 people, and it will take some days to do it. After 6 days of work, 9 people left the team. As a result, it takes 4 more days to complete the project than originally planned. In how many days did they originally plan to finish the project?**

Answer

22

Solution 1

The original team size is 45 people, after 9 leave, the smaller team to the larger team has ratio
$$36 : 45 = 4 : 5.$$
Therefore the ratio of time it takes the two team sizes to complete the same amount of work is $5 : 4$, so the smaller team takes 5 days to complete the work the full team can complete in 4. This means that every 5 days the smaller team works they fall 1 day behind schedule. Therefore in $5 \times 4 = 20$ days they fall 4 days behind schedule. Hence the smaller team takes 20 days to finish the project, for a total of $6 + 20 = 26$ days. Therefore the original plan had the project taking $26 - 4 = 22$ days.

(Algebra) Let x be the amount of work the full team of 45 can do in a single day. Therefore, the original plan is that the project takes $\dfrac{1}{x}$ days to complete. In the 6 days working at full strength, the team finishes $6x$ of the project. After the 9 people leave, it takes the remaining 36 people 4 extra days to complete the project, a total of

$$\frac{1}{x} - 6 + 4 = \frac{1}{x} - 2$$

days. The smaller team can complete

$$\frac{36}{45}x = \frac{4x}{5}$$

of the project in a day. Since the full team works for 6 days and the smaller team works for $\dfrac{1}{x} - 2$ days to complete the project, we have

$$6x + \left(\frac{1}{x} - 2\right) \times \frac{4x}{5} = 1.$$

Distributing the $\dfrac{4x}{5}$ we have

$$6x + \frac{4}{5} - \frac{8x}{5} = 1$$

and combining like terms gives us

$$\frac{22x}{5} = \frac{1}{5}$$

so

$$x = \frac{1}{22}.$$

Therefore the original plan had the project taking $\dfrac{1}{x} = 22$ days.

www.ingramcontent.com/pod-product-compliance
Lightning Source LLC
Chambersburg PA
CBHW080514220326
41599CB00032B/6078